Studies in Computational Intelligence

Volume 550

Series editor

J. Kacprzyk, Warsaw, Poland

For further volumes:
http://www.springer.com/series/7092

About this Series

The series "Studies in Computational Intelligence" (SCI) publishes new developments and advances in the various areas of computational intelligence—quickly and with a high quality. The intent is to cover the theory, applications, and design methods of computational intelligence, as embedded in the fields of engineering, computer science, physics and life sciences, as well as the methodologies behind them. The series contains monographs, lecture notes and edited volumes in computational intelligence spanning the areas of neural networks, connectionist systems, genetic algorithms, evolutionary computation, artificial intelligence, cellular automata, self-organizing systems, soft computing, fuzzy systems, and hybrid intelligent systems. Of particular value to both the contributors and the readership are the short publication timeframe and the world-wide distribution, which enable both wide and rapid dissemination of research output.

Mousmita Sarma · Kandarpa Kumar Sarma

Phoneme-Based Speech Segmentation Using Hybrid Soft Computing Framework

Mousmita Sarma
Department of Electronics
and Communication Engineering
Gauhati University
Guwahati, Assam
India

Kandarpa Kumar Sarma
Department of Electronics
and Communication Technology
Gauhati University
Guwahati, Assam
India

ISSN 1860-949X ISSN 1860-9503 (electronic)
ISBN 978-81-322-3515-6 ISBN 978-81-322-1862-3 (eBook)
DOI 10.1007/978-81-322-1862-3
Springer New Delhi Heidelberg New York Dordrecht London

Printed on acid-free paper

Springer is part of Springer Science+Business Media (www.springer.com)

This work is dedicated to all the researchers of Speech Processing and related technology

Preface

Speech is a naturally occuring nonstationary signal essential not only for person-to-person communication but has become an important aspect of Human Computer Interaction (HCI). Some of the issues related to analysis and design of speech-based applications for HCI have received widespread attention. With continuous upgradation of processing techniques, treatment of speech signals and related analysis from varied angles has become a critical research domain. It is more so with cases where there are regional linguistic orientations with cultural and dialectal elements. It has enabled technologists to visualize innovative applications. This work is an attempt to treat speech recognition with soft computing tools oriented toward a language like Assamese spoken mostly in the northeastern part of India with rich linguistic and phonetic diversity. The regional and phonetic variety observed in Assamese makes it a sound area for research involving ever-changing approaches in speech and speaker recognition. The contents included in this compilation are outcomes of the research carried out over the last few years with emphasis on the design of a soft computing framework for phoneme segmentation used for speech recognition. Though the work uses Assamese as an application language, the concepts outlined, systems formulated, and the results reported are equally relevant to any other language. It makes the proposed framework a universal system suitable for application to soft computing-based speech segmentation algorithm design and implementation.

Chapter 1 provides basic notions related to speech and its generation. This treatment is general in nature and is expected to provide the background necessary for such a work. The contents included in this chapter also provide the necessary motivation, certain aspects of phoneme segmentation, a review of the reported literature, application of Artificial Neural Network (ANN) as a speech processing tool, and certain related issues. This content should help the reader to have a rudimentary familiarization about the subsequent issues highlighted in the work.

Speech recognition research is interdisciplinary in nature, drawing upon work in fields as diverse as biology, computer science, electrical engineering, linguistics, mathematics, physics, and psychology. Some of the basic issues related to speech processing are summarized in Chap. 2. The related theories on speech perception and spoken word recognition model have been covered in this chapter. As ANN is the most critical element of the book, certain essential features necessary for the subsequent portion of the work constitute Chap. 3. The primary topologies covered include the Multi Layer Perceptron (MLP), Recurrent Neural Network (RNN), Probabilistic Neural Network (PNN), Learning Vector Quantization (LVQ), and Self-Organizing Map (SOM). The descriptions included are designed in such a manner that it serves as a supporting material for the subsequent content.

Chapter 4 primarily discusses about Assamese language and its phonemical characteristics. Assamese is an Indo-Aryan language originated from the Vedic dialects and has strong links to Sanskrit, the ancient language of the Indian subcontinent. However, its vocabulary, phonology, and grammar have substantially been influenced by the original inhabitants of Assam such as the Bodos and Kacharis. Retaining certain features of its parent Indo–European family, it has many unique phonological characteristics. There is a host of phonological uniqueness in Assamese pronunciation which shows variations when spoken by people of different regions of the state. This makes Assamese speech unique and hence requires a study exclusively to directly develop a language-specific speech recognition/speaker identification system.

In Chap. 5, a brief overview derived out of a detailed survey of speech recognition works reported from different groups all over the globe in the last two decades is given. Robustness of speech recognition systems toward language variation is a recent trend of research in speech recognition technology. To develop a system that can communicate between human beings in any language like any other human being is the foremost requirement of any speech recognition system. The related efforts in this direction are summarized in this chapter.

Chapter 6 includes certain experimental work carried out. The chapter provides a description of a SOM-based segmentation technique and explains how it can be used to segment the initial phoneme from some Consonant–Vowel–Consonant (CVC) type Assamese word. The work provides a comparison of the proposed SOM-based technique with the conventional Discrete Wavelet Transform (DWT)-based speech segmentation technique. The contents include a description of an ANN approach to speech segmentation by extracting the weight vectors obtained from SOM trained with the LP coefficients of digitized samples of speech to be segmented. The results obtained are better than those reported earlier.

Chapter 7 provides a description of the proposed spoken word recognition model, where a set of word candidates are activated at first on the basis of phoneme family to whichits initial phoneme belongs. The phonemical structure of every natural language provides some phonemical groups for both vowel and consonant phonemes each having distinctive features. This work provides an approach to CVC-type Assamese spoken words recognition by taking advantages of such phonemical groups of Assamese language, where all words of the recognition vocabulary are initially classified into six distinct phoneme families and then the constituent vowel and consonant phonemes are classified within the group. A hybrid framework, using four different ANN structures, is constituted for this word recognition model, to recognize phoneme family and phonemes and thus the word at various levels of the algorithm.

A technique to remove the CVC-type word limitation observed in case of spoken word recognition model described in Chap. 7 is proposed in Chap. 8.

This technique is based on a phoneme count determination block based on K-means Clustering (KMC) of speech data. A KMC algorithm-based technique provides prior knowledge about the possible number of phonemes in a word. The KMC-based approach enables proper counting of phonemes which improves the system to include words with multiple number of phonemes.

Chapter 9 presents a neural model for speaker identification using speaker-specific information extracted from vowel sounds. The vowel sound is segmented out from words spoken by the speaker to be identified. Vowel sounds occur in a speech more frequently and with higher energy. Therefore, situations where acoustic information is noise corrupted vowel sounds can be used to extract different amounts of speaker discriminative information. The model explained here uses a neural framework formed with PNN and LVQ where the proposed SOM-based vowel segmentation technique is used. The speaker-specific glottal source information is initially extracted using LP residual. Later, Empirical Mode Decomposition (EMD) of the speech signal is performed to extract the residual. The work shows a comparison of effectiveness between these two residual features.

The key features of the work have been summarized in Chap. 10. It also includes certain future directions that can be considered as part of follow-up research to make the proposed system a fail-proof framework.

The authors are thankful to the acquisition, editorial, and production team of the publishers. The authors are thankful to students, research scholars, faculty members of Gauhati University, and IIT Guwahati for being connected in respective ways to the work. The authors are also thankful to their respective family members for their support and encouragement.

Finally, the authors are thankful to the Almighty.

Guwahati, Assam, India
January 2014

Mousmita Sarma
Kandarpa Kumar Sarma

Acknowledgments

The authors acknowledge the contribution of the following:

- Mr. Prasanta Kumar Sarma of Swadeshy Academy, Guwahati and Mr. Manoranjan Kalita, sub-editor of Assamese daily Amar Axom, Guwahati for their exemplary help in developing rudimentary know-how on linguistic and phonetics.
- Krishna Dutta, Surajit Deka, Arup, Sagarika Bhuyan, Pallabi Talukdar, Amlan J. Das, Mridusmita Sarma, Chayashree Patgiri, Munmi Dutta, Banti Das, Manas J. Bhuyan, Parismita Gogoi, Ashok Mahato, Hemashree Bordoloi, Samarjyoti Saikia, and all other students of Department of Electronics and Communication Technology, Gauhati University, who provided their valuable time during recording the raw speech samples.

Contents

Part II Design Aspects

Acronyms

AANN Autoassociative Neural Networks are feedforward nets trained to produce an approximation of the identity mapping between network inputs and outputs using backpropagation or similar learning procedures

ANN Artificial Neural Networks are nonparametric models inspired by biological central nervous systems. These are particularly similar like the working of the human brain and are capable of machine learning and pattern recognition. Usually, these are represented as systems of interconnected "neurons" that can compute values from inputs by feeding information through the network

ARPA Advanced Research Projects Agency is an agency of the United States Department of Defense responsible for the development of new technologies for use by the military

ASR Automatic Speech Recognition is a technology by which a computer or a machine is made to recognize the speech of a human being

AVQ Adaptive Vector Quantization is a form of vector quantization where the parameters of a quantizer are updated during real time operation based on observed information regarding the statistics of the signal being quantized

BP Back Propagation Algorithm is a common supervised method of training ANN where in reference to a desired output, the network learns from many inputs

BPTT Back Propagation Through Time is a gradient-based technique for training certain types of Recurrent Neural Network (RNN)s

CFRNN Fully Connected RNN is the basic RNN where each neuron unit has a direct connection to every other neuron unit. The Complex Fully Recurrent Neural Network has both global and local feedback paths and process both real and imaginary segments separately which helps in better learning

CRF Conditional Random Field is a class of statistical modeling used for structured prediction taking context (neighboring samples) into account often applied in pattern recognition and machine learning

CSR Continuous Speech Recognition, ASR technology implemented in case continuous speech signal

CV	It is a coda-less open syllable where speech sounds are organized in the sequence of consonant and vowel
CVC	It is a coda-closed syllable where speech sounds are organized in the sequence of consonant, vowel and consonant
DEKF	Decoupled Extended Kalman Filter is a technique used to train Recurrent Neural Network (RNN) with separated blocks of Kalman filter. It is class of RNN training method where modified forms of Kalman filters are employed in decoupled form
DNN	Deep Neural Network is a type of ANN where learning is layer distributed and retains the learning at neuron levels
DRT	Direct Realist Theory is a theory of speech perception proposed by Carol Fowler which claims that the objects of speech perception are articulatory rather than acoustic events
DTW	Dynamic Time Warping is an algorithm to measure similarity between two temporally varying time sequence
DWFB	Discrete Wavelet Filter Bank is a wavelet filter made up of successive high pass and low pass orthogonal filter derived from the process of wavelet decomposition tree where the time scale representation of the signal to be analyzed is passed through filters with different cutoff frequencies at different scales
DWT	Discrete Wavelet Transform is a kind of wavelet transform where the wavelets are discretely sampled
EKF	Extended Kalman Filter is a nonlinear version of the Kalman filter used in estimation theory which linearizes about an estimate of the current mean and covariance
EMD	Empirical Mode Decomposition is a signal decomposition technique used in case of nonstationary signals derived by N. E. Huang, where the signal is decomposed into a set of high frequency oscillating components and a low frequency residue
F1	The first spectral peak observed in the vocal tract filter response
FF	Feed Forward is an ANN structure where the processing flow is in the forward direction
FFT	Fast Fourier Transform is a computationally efficient algorithm used for Discrete Fourier Transform (DFT) analysis
FVQ	Fuzzy Vector Quantization is a technique where fuzzy concepts are used for vector quantization and clustering
GMM	Gaussian Mixture Model is a probabilistic model for representing the presence of subpopulations within an overall population, without requiring that an observed data set should identify the sub-population to which an individual observation belongs where mixture distribution is Gaussian
HCI	Human Computer Interface involves the study, planning, and design of the interaction between people (users) and computers

HMM Hidden Markov Models is a statistical Markov model in which the system being modeled is assumed to be a Markov process with unobserved or hidden states. It can be considered to be the simplest dynamic Bayesian network

IMF Intrinsic Mode Function is a high frequency component obtained after decomposition using EMD technique from nonstationary signal

ISR Isolated Speech Recognition is an ASR technology developed for isolated speech where a distinct pause is observed between each spoken word

KMC K-means clustering is a method of vector quantization originally from signal processing, that is popular for cluster analysis in data mining

KNN K Nearest Neighbor is a nonparametric method for classification and regression used in pattern recognition that predicts class memberships based on the most common amongst its k closest training examples in the feature space

LD Log Determinant is a function from the set of symmetric matrices which provides a measure of the volume of an ellipsoid precisely

LDA Linear Discriminant Analysis is a method used in statistics, pattern recognition, and machine learning to find a linear combination of features which characterizes or separates two or more classes of objects

LPC Linear Prediction Coding is a tool used mostly in audio signal processing and speech processing for representing the spectral envelope of digital signal of speech in compressed form, using the information of a linear predictive model

LPCC Linear Prediction Cepstral Coefficients are the coefficients that can be found by converting the Linear Prediction coefficients into cepstral coefficients

LVQ Learning Vector Quantization is a prototype based supervised classification algorithm and is the supervised counterpart of vector quantization systems

MBE Minimum Boundary Error is a criterion related to HMM based ASR, which tries to minimize the expected boundary errors over a set of possible phonetic alignments

MFCC Mel-Frequency Cepstral Coefficients are coefficients that collectively make up a Mel-Frequency Cepstrum which is a representation of the short-term power spectrum of a sound, based on a linear cosine transform of a log power spectrum on a nonlinear mel scale of frequency

MLP Multi Layer Perceptron is a feedforward artificial neural network model that maps sets of input data onto a set of appropriate outputs

MMI Maximum Mutual Information is a discriminative training criteria used in the HMM modeling involved in ASR technology, which considers all the classes simultaneously, during training and parameters of the correct model are updated to enhance its contribution to the observations, while parameters of the alternative models are updated to reduce their contributions

MPE Minimum Phone Error is a discriminative criteria called MPE is a smoothed measure of phone transcription error while training HMM parameters. It is introduced by Povey in his Doctoral thesis submitted to University of Cambridge in 2003

MSE Mean Square Error is one of many ways to quantify the difference between values implied by an estimator and the true values of the quantity being estimated

NARX Nonlinear Autoregressive with Exogenous Input is an architectural approach of RNN with embedded memory provides the ability to track nonlinear time varying signals

NWHF Normalized Wavelet Hybrid Feature is a hybrid feature extraction method which uses the combination of Classical Wavelet Decomposition (CWD) and Wavelet Packet Decomposition (WPD) along with z-score normalization technique

PCA Principal Component Analysis is a mathematical procedure that uses orthogonal transformation to convert a set of observations of possibly correlated variables into a set of values of linearly uncorrelated variables called principal components

PDP Parallel Distributed Processing is a class of neurally inspired information processing models that attempt to model information processing the way it actually takes place in the brain, where the representation of information is distributed and learning can occur with gradual changes in connection strength by experience

PLP Perceptual Linear Predictive analysis is a technique of speech analysis which uses the concepts of the critical-band spectral resolution, the equal loudness curve and the intensity-loudness power law of psychophysics of hearing to approximate the auditory spectrum by an autoregressive allpole model

PNN Probabilistic Neural Network is a feedforward neural network, which was derived from the Bayesian network and a statistical algorithm called Kernel Fisher discriminant analysis

PSO Particle Swarm Optimization is a computational method that optimizes a problem by iteratively trying to improve a candidate solution with regard to a given measure of quality

RBF Radial Basis Function is a hybrid supervised–unsupervised ANN topology having a static Gaussian function as the nonlinearity for the hidden layer processing elements

RBPNN Radial Basis Probabilistic Neural Networks is a feedforward ANN model that avoids the huge amount of hidden units of the PNNs and reduce the training time for the RBFs

RD Rate Distortion is a theory related to lossy data compression which addresses the problem of determining the minimal number of bits per symbol, as measured by the rate R, that should be communicated over a channel, so that the source can be approximately reconstructed at the receiver without exceeding a given distortion D

RNN Recurrent Neural Network is a class of ANN where connections between units form a directed cycle which creates an internal state of the network which allows it to exhibit dynamic temporal behavior. It is an ANN with feed forward and feedback paths

RTRL Real Time Recurrent Learning is a training method associated with RNN where adjustments are made to the synaptic weights of a fully connected recurrent network in real time

SGMM Subspace Gaussian Mixture Model is an acoustic model for speech recognition in which all phonetic states share a common Gaussian Mixture Model structure, and the means and mixture weights vary in a subspace of the total parameter space

SOM Self-Organizing Map is a type of ANN that is trained using unsupervised learning to produce a low-dimensional (typically two-dimensional), discretized representation of the input space of the training samples, called a map

SSE Sum of Squares of Error is the sum of the squared differences between each observation and its group's mean

SUR Speech Understanding Research is a program founded by Advanced Research Projects Agency (ARPA) of the U.S. Department of Defense during 1970s

SVM Support Vector Machines are supervised learning models with associated learning algorithms that analyze data and recognize patterns, used for classification and regression analysis

TDNN Time Delay Neural Network is an ANN architecture whose primary purpose is to work on continuous data

VE Voting Experts is an algorithm for chunking sequences of symbols which greedily searches for sequences that match an information-theoretic signature: low entropy internally and high entropy at the boundaries

VOT Voice Onset Time is defined as the length of time that passes between the release of a stop consonant and the onset of voicing, i.e., the vibration of the vocal folds or periodicity

VQ Vector Quantization is a classical quantization technique originally used for data compression which allows the modeling of probability density functions by the distribution of prototype vectors

WER Word Error Rate is a measure of speech recognition system derived from the Levenshtein distance, working at the word level

WPD Wavelet Packet Decomposition is a wavelet transform where the discrete-time (sampled) signal is passed through more filters than the discrete wavelet transform (DWT)

Part I
Background

The book is constituted by two parts. In Part I, we include the background of the work and motivation behind the work. We also discuss some of the relevant literature in circulation which provides an insight into the development of technology related to speech recognition. We also include a discussion on phoneme and its importance in case of speech recognition with certain stress on Assamese language which is widely spoken in north–eastern part of India. We also discuss about ANN as tool of speech recognition and highlight its importance for such an application. We also summarize the contribution of the work and provide a brief organization of the chapters that constitute the work.

Chapter 1
Introduction

Abstract Speech is a naturally occuring non-stationary signal essential not only for person-to-person communication but also is an important aspect of human–computer interaction (HCI). Some of the issues related to analysis and design of speech-based applications for HCI have received widespread attention. Some of these issues are covered in this chapter which is used as background and motivation for the work included in the subsequent portion of the book.

Keywords Speech · Artificial neural network · Phoneme · Segmentation · RNN · SOM

1.1 Background

Speech is the most common medium of communication between person to person. Speech is a naturally occuring non-stationary signal produced from a time-varying vocal tract system. It results due to time-varying excitation. The speech production mechanism is started at the cortex of the human brain which sends some neural signals through the nerves and the muscles which control breathing to enable humans to articulate the word or words needed to communicate the thoughts. This production of the speech signal occurs at physiological level and it involves rapid, coordinated, sequential movements of the vocal apparatus. The purpose of speech is information exchange between human beings. In terms of information theory as introduced by Shannon, speech can be represented in terms of its message content or information. Speech in modern times has become an important element of human–computer interaction (HCI). As such, all speech processing applications are, therefore, oriented toward establishing some connection with HCI-based designs. While doing so, speech needs detailed analysis and processing with proper mechanisms to develop applications suitable for HCI designs.

M. Sarma and K. K. Sarma, *Phoneme-Based Speech Segmentation Using Hybrid Soft Computing Framework*, Studies in Computational Intelligence 550,
DOI: 10.1007/978-81-322-1862-3_1,

An alternative way of characterizing speech is in terms of the signal carrying the message information, i.e., the acoustic waveform [1]. Speech signal contains three distinct events—silence, voiced, and unvoiced. The silence is that part of the signal when no speech is produced. Voiced is that periodic part of the speech signal which is produced due to the periodic vibrations of the vocal tract due to flow of air from the lungs, whereas unvoiced part of the signal is produced due to the non-vibrating vocal cords and hence the resulting signal is aperiodic and random in nature. These three parts of the signal occupy distinct spectral regions in the spectrogram.

However, the information that is communicated through speech is intrinsically of a discrete nature. Hence, it can be represented by a concatenation of a number of linguistically varying sound units called the phonemes, which is possibly the smallest unit of speech signal. Each language has its own distinctive set of phonemes, typically numbering between 30 and 50. Segmentation of phoneme or identification of the phoneme boundary is a fundamental and crucial task since it has many important applications in speech and audio processing. The work described throughout this book presents a new approach to phoneme-level speech segmentation based on a hybrid soft computational framework. A Self-Organizing Map (SOM) trained with various iteration numbers is used to extract out a number of abstract internal structures in terms of weight vector, from any Assamese spoken word in an unsupervised manner. The SOM is an Artificial Neural Network (ANN) trained by following a competitive learning algorithm [2]. In other words, SOM provides some phoneme boundaries, from which some supervised Probabilistic Neural Network (PNN) [2] identifies the constituent phoneme of the word. The PNNs are trained to learn patterns of all Assamese phonemes using a database of clean Assamese phonemes recorded from five male and five female speakers in noise-free and noisy environment. An important aspect of the segmentation and classification algorithm is that it uses the priori knowledge of first formant frequency (F1) of the Assamese phonemes while taking decision, since phonemes are distinguished by its own unique pattern as well as in terms of their formant frequencies. The work uses the concept of pole or formant location determination from the linear prediction (LP) model of vocal tract while estimating F1 [3]. Assamese words containing all the phonemes are recorded from a set of girl and boy speaker, so that the classification algorithm can remove the speaker dependence limitation.

The proposed SOM- and PNN-based algorithm is able to segment out all the vowel and consonant phonemes from any consonant–vowel–consonant (CVC)-type Assamese word. Thus, the algorithm reflects two different application possibilities. The recognized consonant and vowel phonemes can be used to recognize the CVC-type Assamese words, whereas the segmented vowel phonemes can be used for Assamese speaker identification. Some experiments have been performed as part of this ongoing work to explore such future application possibilities of the said algorithm in the field of spoken word recognition and speaker identification.

The phonemical structure of every language provides various phonemical groups for both vowel and consonant phonemes each having distinctive features. A spoken word recognition model can be developed in Assamese language by taking advantage of such phonemical groups. Assamese language has six distinct phoneme families.

Another important fact is that the initial phoneme of a word is used to activate words starting with that phoneme in spoken word recognition models. Therefore, by investigating the initial phoneme, one can classify them into a phonetic group, and then, it can be classified within the group. The second part of this work provides a prototype model for Assamese CVC-type words recognition, where all words of the recognition vocabulary are initially classified into six distinct phoneme families, and then, the constituent vowel and consonant phonemes are segmented and recognized by means of the proposed SOM- and PNN-based segmentation and recognition algorithm. Before using the global decision taken by the PNN, a recurrent neural network (RNN) takes some local decisions about the incoming word and classifies them into six phoneme families of Assamese language on the basis of the first formant frequency (F1) of its neighboring vowel. Then, the segmentation and recognition is performed separately at the RNN decided family. While taking decision about the last phoneme, the algorithm is assisted by some learning vector quantization (LVQ) code book which contains a distinct entry for every word of the recognition vocabulary. It helps the algorithm to take the most likely decision.

The spoken word recognition model explained in this work has the severe limitation of recognizing only CVC-type words. Therefore, in order to include multiple combinations of words in the present method, a logic needs to be developed so that the algorithm can initially take some decision about the number of phonemes in the word to be recognized. As a part of this work, we have proposed a method to take priory decision to determine number of phonemes by using k-mean clustering (KMC) technique.

Vowel phoneme is a part of any acoustic speech signal. Vowel sounds occur in a speech more frequently and with higher energy. Therefore, vowel phoneme can be used to extract different amounts of speaker discriminative information in situations where acoustic information is noise corrupted. Use of vowel sound as the basis for speaker identification has been initiated long back by the Speech Processing Group, University of Auckland, New Zealand [4]. Since then, phoneme recognition algorithms and related techniques have received considerable attention in the domain of speaker recognition and have even been extended to the linguistic arena. Role of vowel phoneme is yet an open issue in the field of speaker verification or identification. This is because of the fact that vowel phoneme-based pronunciation varies with regional and linguistic diversity. Hence, segmented vowel speech slices can be used to track regional variation in the way the speaker speaks the language. It is more so in case of a language like Assamese spoken by over three million people in the northeast (NE) state of Assam with huge linguistic and cultural diversity which have influenced the way people speak the language. Therefore, an efficient vowel segmentation technique shall be effective in speaker identification system. The third part of this work experimented a vowel phoneme segmentation-based speaker identification technique. A separate database of vowel phonemes is created by some samples obtained from Assamese speakers. This clean vowel database is used to design a LVQ-based code book by means of LP residue and empirical mode decomposition (EMD) residue. The LP error sequence provides the speaker source information by subtracting the vocal tract effect, and therefore, it can be used as an effective feature for speaker recognition. Another new method EMD is used to extract residual content from speech signal. EMD is an adaptive tool to analyze nonlinear or non-stationary signals, which segregates the constituent parts of the signal based on the local behavior of the signal. Using EMD, signals can be decomposed into number of frequency modes called intrinsic mode function (IMF) and a residue [5]. Thus, the LVQ code book contains some unique entries for all the speakers in terms of vowel sound's source pattern. Speaker identification is carried out by first segmenting the vowel sound from the speaker's speech signal with the proposed SOM-based segmentation technique and then matching the vowel pattern with the LVQ code book.

A major part of the work is related to phoneme boundary segmentation. In the next section, we provide a survey of certain techniques related to phoneme boundary segmentation which are found to be closely related to the present work.

1.2 Phoneme Boundary Segmentation: Present Technique

Automatic segmentation of speech signals into phonemes is necessary for the purpose of the recognition and synthesis so that manpower and time required in manual segmentation can be removed. Mostly two types of segmentation techniques are found in the literature. These are implicit segmentation and explicit segmentation [6]. Implicit segmentation methods split up the utterance into segments without explicit information like phonetic transcription. It defines segment implicitly as a spectrally stable part of the signal. Explicit segmentation methods split up the incoming utterance into segments that are explicitly defined by the phonetic transcription. Both the methods have their respective advantages and disadvantages. In explicit segmentation, the segment boundaries may be inaccurate due to a possible poor resemblance between reference and test spectrum since the segments are labeled in accordance with the phonetic transcription. Further segmentation can be done on phone level, syllable level, and word level. Here we present a brief literature review on various speech segmentation methods.

1. In 1991, Hemert reported a work [6] on automatic segmentation of speech, where the author combined both an implicit and an explicit method to improve the final segmentation results. First, an implicit segmentation algorithm splits up the utterance into segments on the basis of the degree of similarity between the frequency spectra of neighboring frames. Secondly, an explicit algorithm does the same, but this time on the basis of the degree of similarity between the frequency spectra of the frames in the utterance and reference spectra. A combination algorithm compares the two segmentation results and produces the final segmentation.
2. Deng et al. reported another work on using wavelet probing for compression-based segmentation in 1993, where the authors described how wavelets can be used for data segmentation. The basic idea is to split the data into smooth

segments that can be compressed separately. A fast algorithm that uses wavelets on closed sets and wavelet probing is described in this chapter [7].

3. In 1994, in a work reported by Tang et al. in [8], the design of a hearing aid device based on wavelet transform is explained. The fast wavelet transform is used in the work to decompose speech into different frequency components.
4. Wendt et al. reported a work on pitch determination and speech segmentation using discrete wavelet transform (DWT) in 1996. They have proposed a time-based event detection method for finding the pitch period of a speech signal. Based on the DWT, it detects voiced speech, which is local in frequency, and determines the pitch period. This method is computationally inexpensive, and through simulations and real speech experiments the authors show that it is both accurate and robust to noise [9].
5. Another work is reported by Suh and Lee in [10], where the authors proposed a new method of phoneme segmentation using multilayer perceptron (MLP) which is a feedforward (FF) ANN. The structure of the proposed segmenter consists of three parts: preprocessor, MLP-based phoneme segmenter, and postprocessor. The preprocessor utilizes a sequence of 44 order feature parameters for each frame of speech, based on the acoustic–phonetic knowledge.
6. An automatic method is described for delineating the temporal boundaries of syllabic units in continuous speech using a temporal flow model (TFM) and modulation-filtered spectral features by Shastri et al. [11]. The TFM is an ANN architecture that supports arbitrary connectivity across layers, provides for feed-forward (FF) as well as recurrent links, and allows variable propagation delays along links. They have developed two TFM configurations, global and tonotopic, and trained on a phonetically transcribed corpus of telephone and address numbers spoken over the telephone by several hundred individuals of variable dialect, age, and gender. The networks reliably detected the boundaries of syllabic entities with an accuracy of 84 %.
7. In 2002, Gomez and Castro proposed an work on automatic segmentation of speech at the phonetic level [12]. Here, the phonetic boundaries are established using a dynamic time warping (DTW) algorithm that uses the aposteriori probabilities of each phonetic unit given the acoustic frame. These aposteriori probabilities are calculated by combining the probabilities of acoustic classes which are obtained from a clustering procedure on the feature space and the conditional probabilities of each acoustic class with respect to each phonetic unit. The usefulness of the approach presented in the work is that manually segmented data are not needed in order to train acoustic models.
8. Nagarajan et al. [13] reported a minimum phase group delay-based approach to segment spontaneous speech into syllable-like units. Here, three different minimum phase signals are derived from the short-term energy functions of three subbands of speech signals, as if it were a magnitude spectrum. The experiments are carried out on Switchboard and OGI Multi-language Telephone Speech corpus, and the error in segmentation is found to be utmost 40 ms for 85 % of the syllable segments.

9. Zioko et al. have reported a work where they applied the DWT to speech signals and analyze the resulting power spectrum and its derivatives to locate candidates for the boundaries of phonemes in continuous speech. They compare the results with hand segmentation and constant segmentation over a number of words. The method proved to be effective for finding most phoneme boundaries. The work was published in 2006 [14].
10. In 2006, Awais et al. reported another work [15] where the authors described a phoneme segmentation algorithm that uses fast Fourier transform (FFT) spectrogram. The algorithm has been implemented and tested for utterances of continuous Arabic speech of 10 male speakers that contain almost 2,346 phonemes in total. The recognition system determines the phoneme boundaries and identifies them as pauses, vowels, and consonants. The system uses intensity and phoneme duration for separating pauses from consonants. Intensity in particular is used to detect two specific consonants (*/r/*, */h/*) when they are not detected through the spectrographic information.
11. In 2006, Huggins-Daines et al. reported another work [16], where the authors described an extension to the Baum-Welch algorithm for training Hidden Markov Model (HMM)s that uses explicit phoneme segmentation to constrain the forward and backward lattice. The HMMs trained with this algorithm can be shown to improve the accuracy of automatic phoneme segmentation.
12. Zibert et al. reported a work [17] in 2006, where they have proposed a new, high-level representation of audio signals based on phoneme recognition features suitable for speech and non-speech discrimination tasks. The authors developed a representation where just one model per class is used in the segmentation process. For this purpose, four measures based on consonant–vowel pairs obtained from different phoneme speech recognizers are introduced and applied in two different segmentation–classification frameworks. The segmentation systems were evaluated on different broadcast news databases. The evaluation results indicate that the proposed phoneme recognition features are better than the standard mel-frequency cepstral coefficients and posterior probability-based features (entropy and dynamism). The proposed features proved to be more robust and less sensitive to different training and unforeseen conditions. They have claimed that the most suitable method for speech/non-speech segmentation is a combination of low-level acoustic and high-level recognition features which they have derived by performing additional experiments with fusion models based on cepstral and the proposed phoneme recognition features.
13. An improved HMM/support vector machine (SVM) method for a two-stage phoneme segmentation framework, which attempts to imitate the human phoneme segmentation process, is described by Kuo et al. [18]. The first stage of the method performs HMM forced alignment according to the minimum boundary error (MBE) criterion. The objective of the work was to align a phoneme sequence of a speech utterance with its acoustic signal counterpart based on MBE-trained HMMs and explicit phoneme duration models. The second stage uses the SVM method to refine the hypothesized phoneme boundaries derived by HMM-based forced alignment [18].

14. Another work [19] is reported by Almpanidis and Kotropoulos [19] on phone-level speech segmentation where the authors employed the Bayesian information criterion corrected for small samples and model speech samples with the generalized Gamma distribution, which offers a more efficient parametric characterization of speech in the frequency domain than the Gaussian distribution. A computationally inexpensive maximum likelihood approach is used for parameter estimation. The proposed adjustments yield significant performance improvement in noisy environment.
15. Qiao [20] reported a work on unsupervised optimal phoneme segmentation. The work formulates the optimal segmentation into a probabilistic framework. Using statistics and information theory analysis, the author developed three optimal objective functions, namely mean square error (MSE), log determinant (LD), and rate distortion (RD). Specially, RD objective function is defined based on information RD theory and can be related to human speech perception mechanisms. To optimize these objective functions, the author used time-constrained agglomerative clustering algorithm. The author also proposed an efficient method to implement the algorithm by using integration functions.
16. In 2008, Qiao et al. reported a work [21] on unsupervised optimal phoneme segmentation which assumes no knowledge on linguistic contents and acoustic models. The work formulated the optimal segmentation problem into a probabilistic framework. Using statistics and information theory analysis, they developed three different objective functions, namely summation of square error (SSE), LD, and RD. RD function is derived from information RD theory and can be related to human signal perception mechanism. They introduced a time-constrained agglomerative clustering algorithm to find the optimal segmentations and proposed an efficient method to implement the algorithm by using integration functions. The experiments are carried out on TIMIT database to compare the above three objective functions, and RD provides the best performance.
17. A text-independent automatic phone segmentation algorithm based on the Bayesian information criterion reported in 2008 by Almpanidis and Kotropoulos [22]. In order to detect the phone boundaries accurately, the authors employ an information criterion corrected for small samples while modeling speech samples with the generalized gamma distribution, which offers a more efficient parametric characterization of speech in the frequency domain than the Gaussian distribution. Using a computationally inexpensive maximum likelihood approach for parameter estimation, they have evaluated the efficiency of the proposed algorithm in M2VTS and NTIMIT datasets and demonstrated that the proposed adjustments yield significant performance improvement in noisy environments.
18. Miller and Stoytchev proposed an algorithm for the unsupervised segmentation of audio speech, based on the voting experts (VE) algorithm in 2008, which was originally designed to segment sequences of discrete tokens into categorical episodes. They demonstrated that the procedure is capable of inducing breaks with an accuracy substantially greater than chance and suggest possible avenues of exploration to further increase the segmentation quality [23].

19. Jurado et al. described a work [24] in 2009 text-independent speech segmentation using an improvement method for identification of phoneme boundaries. The modification is based on the distance calculation and selection of candidates for boundaries. From the calculation of distances among MFCC features of frames, prominent values that show transitions among phonemes identifying a phoneme boundary are generated. The modification improves the segmentation process in English and Spanish corpus.
20. In 2009, Patil et al. reported a work [25] to improve the robustness of phonetic segmentation to accent and style variation with a two-staged approach combining HMM broad-class recognition with acoustic–phonetic knowledge-based refinement. The system is evaluated for phonetic segmentation accuracy in the context of accent and style mismatches with training data.
21. In 2010, Bharathi et al. proposed a novel approach for automatic phoneme segmentation that is by hybridizing best phoneme segmentation algorithms HMM, Gaussian Mixture Model (GMM), and Brandts Generalized Likelihood Ratio (GLR) [26].
22. Ziolko et al. [27] reported a non-uniform speech segmentation method using perceptual wavelet decomposition in 2010, which is used for the localization of phoneme boundaries. They have chosen eleven subbands by applying the mean best basis algorithm. Perceptual scale is used for decomposition of speech via Meyer wavelet in the wavelet packet structure. A real-valued vector representing the digital speech signal is decomposed into phone-like units by placing segment borders according to the result of the multiresolution analysis. The final decision on localization of the boundaries is made by analysis of the energy flows among the decomposition levels.
23. Kalinli reported a work [28] on automatic phoneme segmentation using auditory attention features in 2012. The auditory attention model can successfully detect salient audio sounds in an acoustic scene by capturing changes that make such salient events perceptually different than their neighbors. Therefore, in this work, the author uses it as an effective solution for detecting phoneme boundaries from acoustic signal. The proposed phoneme segmentation method does not require transcription or acoustic models of phonemes.
24. By mimicking human auditory processing, King et al. has reported a work [29] on speech segmentation where phone boundaries are located without prior knowledge of the text of an utterance. A biomimetic model of human auditory processing is used in this work to calculate the neural features of frequency synchrony and average signal level. Frequency synchrony and average signal levels are used as input to a two-layered SVM-based system to detect phone boundaries. Phone boundaries are detected with 87.0 % precision and 84.8 % recall when the automatic segmentation system has no prior knowledge of the phone sequence in the utterance. The work was published in 2013.

1.3 ANN as a Speech Processing and Recognition Tool

Speech is a commonly occurring means of communication between people to people which is acquired naturally. The ability to adopt speech as a mode of interpersonal communication is developed in stages starting from birth of a person and is executed so smoothly that it hides the complexities associated with its generation. The generation of speech and its use for communication involves the human vocal tract articulation of different biological organs under conscious control affected by factors like gender to upbringing to emotional state. As a result, vocalizations are observed with respect to accent, pronunciation, articulation, roughness, nasality, pitch, volume, and speed; moreover, during transmission, the irregular speech patterns can be further distorted by background noise and echoes, as well as electrical characteristics (if telephones or other electronic equipment are used). All these sources of variability make speech recognition, even more than speech generation, a very complex problem. But people are efficient in recognizing speech more than the conventional computing systems. This is because of the fact that while conventional computers are capable of doing number crunching operations, the human brain is expert in doing pattern recognition which is a cognitive process. The brain adopts a connectionist approach of computation relating past knowledge with the present stimulus to arrive at a decision. The cognitive properties demonstrated by the human brain is based on parallel and distributed processing carried out by a network of biological neurons connected by neurochemical synaptic links (synapses) which show modifications with learning and experience, directly supporting the integration of multiple constraints, and providing a distributed form of associative memory. These attributes have motivated the adoption of learning-based systems like ANN for speech recognition applications.

The initial attempts to use ANN is speech recognition was a conscious effort to treat the entire issue as a pattern recognition problem. Since speech is a pattern and ANNs are efficient at recognizing patterns, therefore, the challenge of dealing with speech recognition was simplified to the recognition of speech samples with ANN trained for the purpose. The earliest attempts could be summarized to tasks like classifying speech segments as voiced/unvoiced, or nasal/fricative/plosive. Next, ANNs became popular for phoneme classification as part of larger speech recognition and related applications. Traditionally, ANNs are adopted for speech applications in two broad forms. In the static form, the ANN receives the entire speech sample as single pattern and the decision is just a pattern association problem. A simple but elegant experiment was performed by Huang and Lippmann in 1988, to show the effectiveness of ANN in a rudimentary form to classify speech samples. This attempt used a MLP with only 2 inputs, 50 hidden units, and 10 outputs trained with vowels uttered by men, women, and children with training sessions sustained up to 50,000 iterations. Later, Elman and Zipser et al. in 1987 trained a network to classify the vowels using a more complex setup. Such an approach is satisfactory with phoneme classification but fails with a continuous speech. In such a case, a dynamic approach is taken where a sliding window rides over a speech sample and feeds a portion of the extract to the ANN which is provided with the ability to capture temporal variations in the speech inputs. The ANN provides a series of local decisions which combine together to generate a global output. Using the dynamic approach, Waibel et al. in 1987–1989 demonstrated the effectiveness of a time delay neural network (TDNN)—a variation of the MLP to deal with phoneme classification.

In other attempts, Kohonen's electronic typewriter [30] used the clustering and classification characteristics of the SOM to obtain an ordered map from a sequence of feature vectors. In the TDNN [31] and Kohonen's electronic typewriter options, time warping was a major issue. To integrate large time spans in the ANN-based speech recognition, hybrid approaches have been adopted. Traditionally, in such hybrid forms immediately after the phoneme recognition block, either HMM models [32] or time delay warping (TDW) [33] measure procedures are used which provides better results. Lately, Deep Neural Networks (DNNs) with many hidden layers and trained using new methods have been shown better results than GMMs in speech recognition applications [34]. These are being adopted for acoustic modeling in speech recognition.

1.3.1 Speech Recognition Using RNN

ANNs are learning-based prediction tools which have been considered for a host of pattern classification applications including speech recognition. As shall be discussed later, ANNs learn the patters applied to them, retain the learning, and use it subsequently. One of the most commonly used ANN structures for pattern classification is the MLP which has found application in speech synthesis and recognition [35–37] as well. Initial attempts of using MLP for speech applications revolved around the assumption that speech as a sample should be learnt by the MLP and discriminated. The MLPs work well as an effective classifier for vowel sounds with stationary spectra, while their phoneme discriminating power deteriorates for consonants characterized by variations of short-time spectra [38]. A simple but elegant experiment was performed by Huang and Lippmann in 1988, to establish the effectiveness of ANN in a rudimentary feedforward form to classify speech samples. In this case, the MLP was designed with only 2 inputs, 50 hidden units, and 10 outputs trained with vowels uttered by men, women, and children. The learning involved training sustained up to 50,000 iterations. Later, Elman and Zipser et al. in 1987 configured and trained a network to classify the vowels using a more complex setup. Such an approach is satisfactory with phoneme classification but fails with a continuous speech. As an optimal solution to these situations, a dynamic approach is formulated where a sliding window rides over a speech sample and feeds a portion of the extract to the ANN. The ANN is provided with the ability to capture temporal variations in the speech inputs. The ANN derives a series of local decisions that combine together to generate a global output. Certain variations in the basic MLP-based approach were executed for making it suitable for dynamic situations. Using such a dynamic approach, Waibel et al. in 1987–1989 demonstrated the effectiveness of a

TDNN—a variation of the MLP. It was intended to deal with phoneme classification. The TDNN has memory or delay blocks either at the input or at the output or both. It adds to the temporal processing ability of the TDNN, but computational requirements rise significantly. Programming the TDNN is a time-consuming process and hence is not considered for real-time applications. Computationally as the TDNN is much demanding than the MLP, TDNN-based real-time applications show considerable slow down for receiving favor and hence has been discarded as a viable option. In such a situation, the RNNs emerge as a viable option. Its design enables it to deal with dynamic variations in the input. It has feedforward and feedback back paths which contribute toward the temporal processing ability. The feedforward paths make it like the MLP; hence, it is able to make nonlinear discrimination between boundaries using a gradient descent-based learning. Next, the feedback paths enable the RNN to generate contextual processing. The work of the RNN therefore involves the generation of a combined output of the feedforward and the feedback paths and the information content fed by a state vector representing the current or contextual portion of the sample for which the response is being generated. The key difference compared to the MLP is the contextual processing which circulates the most relevant portion of the information among the different layers of the network and the constituent neurons. Further, in many situations, due to inversion in the applied patterns while performing the contextual processing, differential mode learning in the local level of neurons enables the RNN to consider only the most relevant portion of the data. With different types of activation functions at different layers of the network, contextual and differential processing is strengthened. For example, in a three hidden layer RNN, if one layer has tan-sigmoidal, the next with log-sigmoidal and the last with tan-sigmoidal activation function enables better lettering. The least correlated portion of the patterns is retained and circulated, and the portions with similarity are discarded. As a result, the RNN becomes a fast learner and tracks time-dependent variations. The RNN uses feedforward and feedback paths to track finer variations in the input. The feedback paths are sometimes passed through memory blocks which enable delayed versions of the processed output to be considered for processing. These variations can be due to the time-dependent behavior of the input patterns. So while the MLP is only able to do discriminations between applied samples, the RNN is able to distinguish classes that show time variations. For the above-mentioned attributes, RNNs are found to be suitable for application like speech recognition [37]. RNNs were first applied to speech recognition in [39]. Other important works include [40–47]. In [40], fully connected RNN is used for speech modeling. By using discriminative training, each RNN speech model is adjusted to reduce its distance from the designated speech unit while increasing distances from the others. In [41] and [42], RNN is used for phone recognition. In [43], Elman network is used for phoneme recognition where HMM-based postprocessing is used, whereas in [44] and [45], HMM-RNN hybrid system is explained. In [46], RNN is used explicitly to model the Markovian dynamics of a set of observations through a nonlinear function with a much larger hidden state space than traditional sequence models such as an HMM. Here, pretraining principles used for DNNs and second-order optimization techniques are used to train an RNN model. In [47], a contextual real-valued input vector is provided in association with each RNN-based word model.

1.3.2 Speech Recognition Using SOM

Among various tools developed during the last few decades for pattern recognition purpose, ANNs have received attention primarily for the fact that these not only can learn from examples but can also retain and use the knowledge subsequently. One such approach requires continuous feeding of examples such that the ANN executes the learning in a supervisory mode. A concurrent approach avoids the requirement of continuous feeding of the training examples. This is possible due to the use of unsupervised learning as demonstrated by the SOM. Proposed by Kohonen, SOM has a feedforward structure with a single computational layer of neurons arranged in rows and columns. Each neuron in the input layer is fully connected to all the source units by some connectionist weights and follows a philosophy of self-organization. The objective is to achieve a dimensionality reduction. The unsupervised learning process groups the features from the continuous input space into discrete units based on certain selection criteria. The SOM converts a large set of input vectors by finding a smaller set of prototypes so as to provide a proper approximation to the original input space. This was the basis of using the SOM for design of the phonetic typewriter by Kohonen. The typewriter was designed to work in real time by recognizing phonemes from a continuous speech so that a text equivalent could be generated. The feature set was generated using a combination of filtering and Fourier transforming of data sampled every 9.83 ms from spoken words which produced a set of 16 dimensional spectral vectors. These constituted the input segment to the SOM, while the output was formed by a grid of nodes arranged in 8 by 12 grid structure. Time-sliced waveforms of the speech was used to train the SOM to constitute certain clusters of phonemes. This enabled the SOM to reorganize and form nodes linked to specific phonemes in an ideal mode. The training was carried out in two stages. During the first stage, speech feature vectors were applied into the SOM and a convergence to an ideal state was generated. During the second stage, the nodes of the SOM used certain labeling linked to specific phonemes. It mapped the phoneme features to a certain label assigned to a node of the SOM. The nodes reorganized during training and provided the optimum topology using a combination of clustering and classification methods. At the end of the training, continuous speech samples were applied which produced a string of labels. These label strings were next compared with the ideal case using a Euclidean distance measure to establish the effectiveness of the training. The recognition segment was performed by a combination of HMM and dynamically expanding context (DEC) approach. Sequences of speech with phonemes of typically 40–400 ms length and span of several time slices were used, and certain postprocessing operation was performed to provide optimum results. Later, several works used SOM and other AI tools as combinations to provide satisfactory performance. Some of the relevant works are [48–51].

1.4 Motivation

After a detail literature survey, it becomes clear that segmentation of speech signal into constituent phonemes is the crucial component of most of the application in the field of speech processing. The success rate of the work like speech recognition or phoneme segmentation-based speaker recognition depends on the accuracy generated by the segmentation algorithm to lay phoneme boundary. Therefore, development of a more efficient segmentation technique is yet an open issue in the field of speech processing. As per the literature survey associated with this work, the most frequently used technique is the constant segmentation and DWT based technique [7, 9, 14]. A few researchers use certain HMM- or MLP-based technique [10, 16, 18, 26], whereas very few use FFT-based technique [15]. But constant segmentation has a risk of loosing information since phoneme length never remains constant. Therefore, it has the problem of loosing boundary information, which indirectly leads to low success rate in recognition. Another popular technique of phoneme segmentation is the DWT-based segmentation. But DWT has the severe problem of downsampling at each decomposition level to avoid aliasing and corresponding upsampling at reconstruction level. It ultimately leads to the problem of losing low-frequency information where voiced phonemes occur. This work also initially considered the DWT-based segmentation, but success rate obtained by DWT has not increased beyond 86 %. This has given the basic motivation to adopt the SOM-based segmentation technique. Phonemes are linguistic abstraction with varying length, and SOM has the capability to capture some abstract internal structure from the data provided to it. SOM uses an unsupervised paradigm of learning which follows the human learning method directly.

The SOM-based phoneme segmentation technique used in the work provides better success rate (around 96 %) while employed for CVC-type Assamese spoken word recognition purpose. Initial phoneme is used to activate words starting with that phoneme as suggested in the renowned spoken word recognition models like Cohort in 1976, Trace in 1986, etc. Investigation of initial phoneme is critical in classification into a phonetic group. The phonemical structure of every language provides various phonemical groups for both vowel and consonant phonemes each having distinctive features. Assamese also has such six distinct phonemical groups. This work provides an approach to phoneme recognition by taking advantage of such phonemical groups of Assamese language, where all words of the recognition vocabulary are initially classified into six phoneme families, and then, the phonemes are identified by some sort of segmentation technique.

Further, some experimental work has been carried out to develop a prototype model for vowel-based Assamese speaker identification. Assamese is an phonetically distinct language, widely spoken in the NE state of Assam, which has regional vowel sound variation between speakers. From region to region and speaker to speaker, the language shows some notable variations, most of the time in the occurrence of the vowel sounds, which clearly reflects the importance of designing an Assamese speaker recognition system capable of dealing with such regional variations of occurrence of vowel sound in the same word. Among the present researchers, phoneme recognition algorithms and related techniques have received considerable attention in the problem of speaker recognition. Therefore, necessity of developing a vowel-based speaker identification model for Assamese native speakers is an obvious challenge which has been reported in the chapters constituting the book.

1.5 Contribution

The following contributions are described in the book:

1. Phoneme Segmentation Technique using SOM:
   In this work, a new phoneme segmentation technique is proposed, based on some abstract internal representation obtained from SOM. The detail of the proposed phoneme segmentation technique is explained in Chap. 6.
2. Application of the Proposed Phoneme Segmentation Technique in CVC type Assamese Word Recognition:
   The proposed phoneme segmentation technique is used in recognizing CVC type word of Assamese language. The details of the proposed spoken word recognition technique is explained in Chap. 7. Application of cluster techniques to generate apriori knowledge for this spoken word recognition model is described in Chap. 8.
3. Vowel Phoneme Segmentation based Assamese Speaker Identification:
   Assamese is a language which shows some distinctive variation in the occurrence of vowel sound in the same word spoken by speaker of different dialect. Chapter 9 describes application of the Proposed Phoneme Segmentation Technique in Vowel based Assamese Speaker Identification using LP residual and EMD residual as speaker discriminative feature.

1.6 Organization

The material in this book are organized in ten chapters. This chapter provides a brief introduction and motivation of the work described throughout the book. In Chaps. 2 and 3 the basic considerations of Speech Processing and ANN technology is explained. An introduction to the phonemical structure of Assamese language is given in Chap. 4. Chapter 5 provides a brief review on recent speech recognition technology. The proposed SOM based phoneme segmentation algorithm is explained in Chap. 6 and a spoken word recognition model using hybrid ANN structure is explained in Chap. 7 where the SOM based phoneme segmentation algorithm is used. In Chap. 8 an algorithm based on KMC to count number of phonemes in a word is explained so that the CVC type word limitation can be removed from the proposed word recognition model. The usefulness of the SOM based segmentation algorithm in vowel based speaker identification is explained in Chap. 9. Finally in

Chap. 10 a conclusion of the work described throughout the book is provided along with some future direction.

References

1. Rabiner LR, Schafer RW (2009) Digital processing of speech signals, 3rd edn. Pearson Education, Dorling Kindersley (India) Pvt. Ltd, Delhi, India
2. Haykin S (2009) Neural network and learning machine, 3rd edn. PHI Learning Private Limited, New Delhi, India
3. Snell RC, Milinazzo F (1993) Formant location from LPC analysis data. IEEE Trans Speech Audio process 1(2):129–134
4. Templeton PD, Guillemin BJ (1990) Speaker identification based on vowel sounds using neural networks. In: Proceedings of 3rd international conference on speech science and technology, Melbourne, Australia, pp 280–285
5. Sheth SS Extraction of pitch from speech signals using Hilbert Huang transform. Available via http://www.iitg.ernet.in/scifac/cep/public_html/Resources/ece_kailash/files/07010245_ExtractionOfPitchFromSpeechSignalsUsingHilbertHuangTransform.pdf
6. Hemert JPV (1991) Automatic segmentation of speech. IEEE Trans Signal Process 39(4):1008–1012
7. Sweldens W, Deng B, Jawerth BD, Peters G (1993) Wavelet probing for compression based segmentation. In: Proceedings of SPIE conference, vol 2034, pp 266–276
8. Tang BT, Lang R, Schroder H (1994) Applying wavelet analysis to speech segmentation and classification. Wavelet applications. In: Proceedings of SPIE, vol 2242
9. Wendt C, Petropulu AP, Peters G (1996) Pitch determination and speech segmentation using the discrete wavelet transform. In: Proceedings of IEEE international symposium on circuits and systems, vol 2
10. Suh Y, Lee Y (1996) Phoneme segmentation of continuous speech using multi-layer perceptron. In: Proceedings of ICSLP
11. Shastri L, Chang S, Greenberg S (1999) Syllable detection and segmentation using temporal flowneural networks. In: Proceedings of the 14th international congress of phonetic sciences, San Francisco
12. Gomez JA, Castro MJ (2002) Automatic segmentation of speech at the phonetic level. In: Structural, syntactic, and statistical pattern recognition. Lecture notes in computer science, vol 2396, pp 672–680
13. Nagarajan T, Murthy HA, Hegde RM (2003) Segmentation of speech into syllable-like units. In: Proceedings of EUROSPEECH, Geneva, Switzerland
14. Zioko B, Manandhar S, Wilson RC (2006) Phoneme segmentation of speech. In: Proceedings of 18th international conference on pattern recognition, vol 4
15. Awais MM, Ahmad W, Masud S, Shamail S (2006) Continuous Arabic speech segmentation using fft spectrogram. In: Proceedings of innovations in information technology, pp 1–6
16. Huggins-Daines D, Rudnicky AI (2006) A constrained Baum-Welch algorithm for improved phoneme segmentation and efficient training. In: Proceedings of interspeech
17. Zibert J, Pavesic N, Mihelic F (2006) Speech/non-speech segmentation based on phoneme recognition features. EURASIP J Appl Signal Process 2006:113
18. Kuo J, Lo H, Wang H (2007) Improved HMM vs. SVM methods for automatic phoneme segmentation. In: Proceedings of INTERSPEECH, pp 2057–2060
19. Almpanidis G, Kotropoulos C (2007) Automatic phonemic segmentation using the Bayesian information criterion with generalised gamma priors. In: Proceedings of 15th European signal processing conference, Poznan, Poland
20. Qiao Y (2008) On unsupervised optimal phoneme segmentation. In: Proceedings of IEEE international conference on acoustics, speech and signal processing, pp 3989–3992

21. Qiao Y, Shimomura N, Minematsu N (2008) Unsupervised optimal phoneme segmentation: objectives, algorithm and comparisons. In: Proceedings of ICASSP, pp 3989–3992
22. Almpanidis G, Kotropoulos C (2008) Phonemic segmentation using the generalised gamma distribution and small sample bayesian information criterion. J Speech Commun 50(1):38–55
23. Miller M, Stoytchev A (2008) Unsupervised audio speech segmentation using the voting experts algorithm. Available via http://research.microsoft.com/en-us/um/people/xiaohe/nips08/paperaccepted/nips2008wsl104.pdf
24. Jurado RS, Gomez-Gil P, Garcia CAR (2009) Speech text-independent segmentation using an improvement method for identification of phoneme boundaries. In: Proceedings of international conference on electrical, communications, and computers, pp 20–24
25. Patil V, Joshi S, Rao P (2009) Improving the robustness of phonetic segmentation to accent and style variation with a two-staged approach. In: Proceedings of the INTERSPEECH, ISCA, pp 2543–2546
26. Bharathi B, Prathiba K (2010) A novel approach for automatic phoneme segmentation. In: Proceedings of international conference on information science and applications, Chennai
27. Zioko M, Gaka J, Zioko B, Drwiega T (2010) Perceptual wavelet decomposition for speech segmentation. In: Proceedings of INTERSPEECH, Makuhari, Chiba, Japan
28. Kalinli O (2012) Automatic phoneme segmentation using auditory attention features. In: Proceedings of INTERSPEECH, ISCA
29. King S, Hasegawa-Johnson M (2013) Accurate speech segmentation by mimicking human auditory processing. In: Proceedings of ICASSP, Vancouvar, Canada
30. Kohonen T (1988) The neural phonetic typewriter. Computer 21(3):11–22
31. Lang KJ, Waibel AH (1990) A time-delay neural network architecture for isolated word recognition. Neural Netw 3:23–43
32. Singer E, Lippmann RP (1992) A speech recognizer using radial basis function neural networks in an HMM framework. In: Proceedings of the IEEE ICASSP
33. Hild H, Waibel A (1993) Multi-speaker/speaker-independent architectures for the multi-state time delay neural network. In: Proceedings of the IEEE ICNN
34. Hinton G, Deng L, Yu D, Dahl GE, Mohamed A, Jaitly N, Senior A, Vanhoucke V, Nguyen P, Sainath TN, Kingsbury B (2012) Deep neural networks for acoustic modeling in speech recognition. IEEE Signal Process Mag 82–97
35. Jurafsky D, Martin JH (2000) Speech and language processing: an introduction to natural language processing, computational linguistics, and speech recognition, 1st edn. Prentice Hall, New Jersey
36. Paul AK, Das D, Kamal M (2009) Bangla speech recognition system using LPC and ANN. In: Proceedings of 7th international conference on advances in pattern recognition, pp 04–06
37. Dede G, Sazl MH (2010) Speech recognition with artificial neural networks. Digit Signal Process 20(3):763–768
38. Ahmad AM, Ismail S, Samaon DF (2004) Recurrent neural network with backpropagation through time for speech recognition. In: Proccedings of international symposium on communications and information technologies, Sapporo, Japan
39. Robinson T, Hochberg M, Renals S (1994) IPA: improved phone modelling with recurrent neural networks. In: Proceedings of IEEE ICASSP
40. Lee T, Ching PC, Chan LW (1995) An RNN based speech recognition system with discriminative training. In: Proceedings of the 4th European conference on speech communication and technology, pp 1667–1670
41. Jamieson LHRC (1996) Experiments on the implementation of recurrent neural networks for speech phone recognition. In: Proceedings of the 30th annual asilomar conference on signals, systems and computers, Pacific Grove, California, Nov 1996, pp 779–782
42. Koizumi T, Mori M, Taniguchi S, Maruya M (1996) Recurrent neural networks for phoneme recognition. In: Proceedings 4th international conference ICSLP 96, vol 1, pp 326–329
43. Rothkrantz LJM, Nollen D (1999) Speech recognition using Elman neural networks. In: Text, speech and dialogue. Lecture notes in computer science, vol 1692, pp 146–151

44. Yan ZX, Yu W, Wei X (2001) Speech recognition model based on recurrent neural networks. Available via http://coitweb.uncc.edu/~ywang32/research/RNN.pdf
45. Sun Y, Bosch LT, Boves L (2010) Hybrid HMM/BLstm-Rnn for robust speech recognition. In: Proceedings of 13th international conference on text, speech and dialogue. Springer, Berlin, Heidelberg, pp 400–407
46. Vinyals O, Ravuri SV, Povey D (2012) Revisiting recurrent neural networks for robust ASR. In: Proceedings of IEEE international conference on acoustics, speech, and signal processing (ICASSP)
47. Mikolov T, Zweig G (2012) Context dependent recurrent neural network language model. In: Proceedings of the IEEE spoken language technology workshop (SLT), Miami, FL, USA, pp 234–239
48. Kasabov N, Nikovski D, Peev E (1993) Speech recognition based on Kohonen self organizing feature maps and hybrid connectionist systems. In: Proceedings of 1st New Zealand international two-stream conference artificial neural networks and expert systems, pp 113–117
49. Beauge L, Durand S, Alexandre F (1993) Plausible self-organizing maps for speech recognition. In: Artificial neural nets and genetic algorithms, pp 221–226
50. Venkateswarlu RLK, Novel RV (2011) Approach for speech recognition by using self organized maps. In: 2011 international conference on emerging trends in networks and computer communications (ETNCC), pp 215–222
51. Kohonen T, Somervuo P (1997) Self-organizing maps of symbol strings with application to speech recognition. In: Proceedings of 1st international workshop on self organizing map, pp 2–7

References

Chapter 2
Speech Processing Technology: Basic Consideration

Abstract In this chapter, we have described the basic considerations of speech processing technology. Starting with the fundamental description of speech recognition systems, we have described the speech communication chain, mechanisms of speech perception and the popular theories, and model of spoken word recognition.

Keywords Spoken word · Speech communication · Perception · Recognition · Hearing

2.1 Fundamentals of Speech Recognition

Speech recognition is special case of speech processing. It deals with the analysis of the linguistic contents of a speech signal. Speech recognition is a method that uses an audio input for data entry to a computer or a digital system in place of a keyboard. In simple terms, it can be a data entry process carried out by speaking into a microphone so that the system is able to interpret the meaning out of it for further modification, processing, or storage. Speech recognition research is interdisciplinary in nature, drawing upon work in fields as diverse as biology, computer science, electrical engineering, linguistics, mathematics, physics, and psychology. Within these disciplines, pertinent work is being done in the areas of acoustics, artificial intelligence, computer algorithms, information theory, linear algebra, linear system theory, pattern recognition, phonetics, physiology, probability theory, signal processing, and syntactic theory [1].

Speech is a natural mode of communication for human. Human start to learn linguistic information in their early childhood, without instruction, and they continue to develop a large vocabulary with the confines of the brain throughout their lives. Humans achieve this skill so naturally that it never facilitates anything specific enabling a person to realize how complex the process is. The human vocal tract and articulators are biological organs with non-linear properties, whose operation is not just under conscious control but also affected by factors ranging from gender to upbringing to emotional state. As a result, vocalizations can vary widely in terms

M. Sarma and K. K. Sarma, *Phoneme-Based Speech Segmentation Using Hybrid Soft Computing Framework*, Studies in Computational Intelligence 550,
DOI: 10.1007/978-81-322-1862-3_2, © Springer India 2014

of their accent, pronunciation, articulation, roughness, nasality, pitch, volume, and speed. Moreover, during transmission, our irregular speech patterns can be further distorted by background noise and echoes, as well as electrical characteristics. All these sources of variability make speech recognition, a very complex problem [2]. Speech recognition systems are generally classified as follows:

1. Continuous Speech Recognition and
2. Discrete Speech Recognition

Discrete systems maintain a separate acoustic model for each word, combination of words, or phrases and are referred to as isolated (word) speech recognition (ISR). Continuous speech recognition (CSR) systems, on the other hand, respond to a user who pronounces words, phrases, or sentences that are in a series or specific order and are dependent on each other, as if linked together.

Discrete speech was used by the early speech recognition researchers. The user has to make a short pause between words as they are dictated. Discrete speech systems are particularly useful for people having difficulties in forming complete phrases in one utterance. The focus on one word at a time is also useful for people with a learning difficulty.

Today, continuous speech recognition systems dominate [3]. However, discrete systems are useful than continuous speech recognition systems in the following situations:

1. For people with speech and language difficulties, it can be very difficult to produce continuous speech. Discrete speech systems are more effective at recognizing nonstandard speech patterns, such as dysarthric speech.
2. Some people with writing difficulties benefit from concentrating on a single word at a time that is implicit in the use of a discrete system. In the continuous speech, people need to be able to speak at least a few words, with no pause, to get a proper level of recognition. Problems can appear if one changes his/her mind about a word, in the middle of a sentence.
3. Children in school need to learn to distinguish between spoken English and written English. In a discrete system, it is easier to do, so since attention is focused on individual words.

However, discrete speech is an unnatural way of speaking, whereas continuous speech systems allow the user to speak with a natural flow at a normal conversational speed [3]. Both discrete and continuous systems may be speaker dependent, independent, or adaptive. Speaker-dependent software works by learning the unique characteristics of a single person's voice, in a way similar to voice recognition. New users must first train the software by speaking to it, so the computer can analyze how the person talks. This often means, users have to read a few pages of text to the computer before they can use the speech recognition software. Conversely, speaker-independent software is designed to recognize anyone's voice, so no training is involved. This means that, it is the only real option for applications such as interactive voice response systems, where businesses cannot ask callers to read pages of text

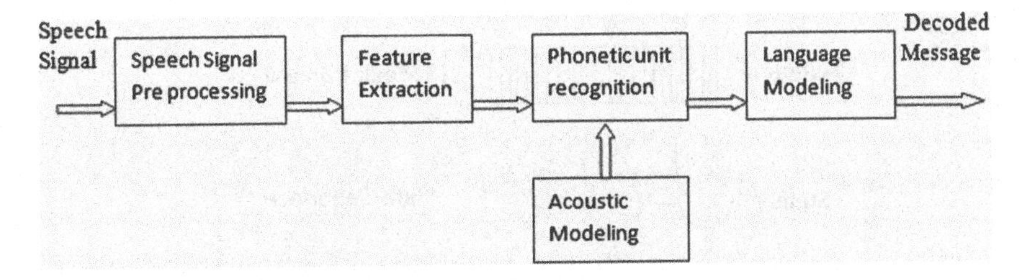

Fig. 2.1 Structure of a standard speech recognition system [2]

before using the system. The other side of the picture is that speaker-independent software is generally less accurate than speaker-dependent software [4].

Speech recognition engines that are speaker independent generally deal with this fact by limiting the grammars they use. By using a smaller list of recognized words, the speech engine is more likely to correctly recognize what a speaker said. Speaker-dependent software is commonly used for dictation software, while speaker-independent software is more commonly found in telephone applications [4].

Modern speech recognition systems also use some adaptive algorithm so that the system becomes adaptive to speaker-varying environment. Adaptation is based on the speaker-independent system with only limited adaptation data. A proper adaptation algorithm should be consistent with speaker-independent parameter estimation criterion and adapt those parameters that are less sensitive to the limited training data [5]. The structure of a standard speech recognition system is illustrated in Fig. 2.1 [2]. The elements of the system can be described as follows:

1. **Raw speech**: Speech is typically sampled at a high frequency, for example, 16 kHz over a microphone or 8 kHz over a telephone. This yields a sequence of amplitude values over time.
2. **Pre-processing**: Prior to applying the speech signal, preprocessing of speech signals is required. Although the speech signals contain all the necessary information within 4 kHz, raw speech signals naturally have very large bandwidth. High-frequency speech signals are always less informative and contains noise. Therefore, to make the signal noiseless, preprocessing involves a digital filtering operation. Various structures of digital filters can be used for these purposes.
3. **Signal analysis**: Raw speech should be initially transformed and compressed, in order to simplify subsequent processing. Many signal analysis techniques are available that can extract useful features and compress the data by a factor of ten without losing any important information.
4. **Speech frames**: The result of signal analysis is a sequence of speech frames, typically at 10 ms intervals, with about 16 coefficients per frame. These frames may be augmented by their own first and/or second derivatives, providing explicit information about speech dynamics. This typically leads to improved performance. The speech frames are used for acoustic analysis [2].

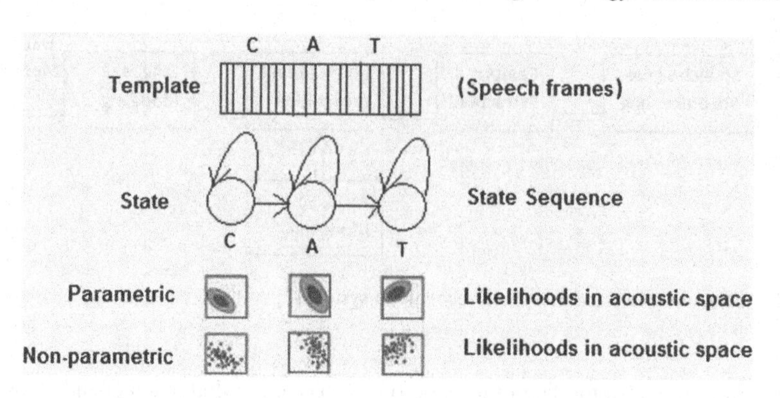

Fig. 2.2 Acoustic models: template and state representations for the word 'cat' [2]

5. **Acoustic models**: In order to analyze the speech frames for their acoustic content, we need a set of acoustic models. There are many kinds of acoustic models, varying in their representation, granularity, context dependence, and other properties. Figure 2.2 shows two popular representations for acoustic models. The simplest is a template, which is just a stored sample of the unit of speech to be modeled, for example, a recording of a word. An unknown word can be recognized by simply comparing it against all known templates and finding the closest match. Templates have some major drawbacks. One is, they cannot model acoustic variabilities, except in a coarse way by assigning multiple templates to each word. Furthermore, they are limited to whole word models, because it is difficult to record or segment a sample shorter than a word. Hence, templates are useful only in small systems.
   A more flexible representation, used in larger systems, is based on trained acoustic models, or states. In this approach, every word is modeled by a sequence of trainable states, and each state indicates the sounds that are likely to be heard in that segment of the word, using a probability distribution over the acoustic space. Probability distributions can be modeled parametrically, by assuming that they have a simple shape like Gaussian distribution and then trying to find the parameters that describe it. Another method of modeling probability distributions is nonparametric, that means, by representing the distribution directly with a histogram over a quantization of the acoustic space or as we shall see with an ANN.

2.2 Speech Communication Chain

The fundamental purpose of generation of speech is communication [6]. Speech communication involves a chain of physical and psychological events. The speaker initiates the chain by setting into motion the vocal apparatus, which propels a modulated airstream to the listener. The process culminates with the listener receiving the

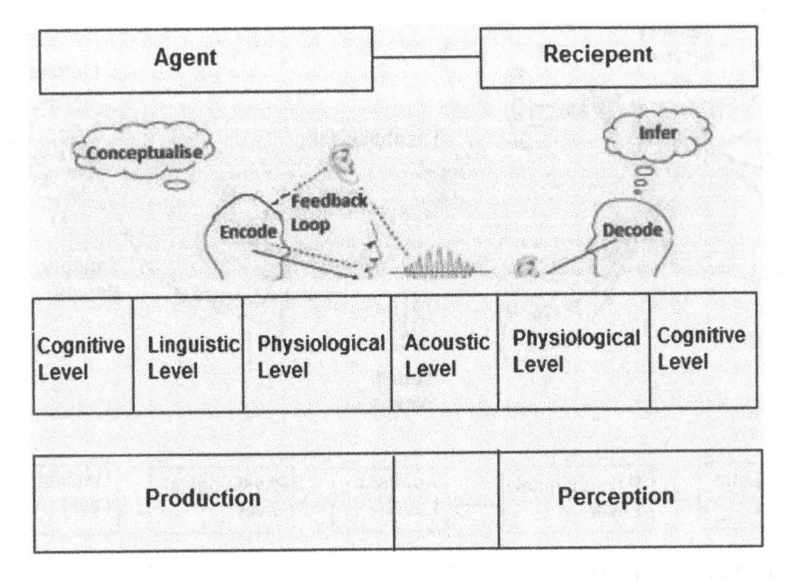

Fig. 2.3 Communication model [8]

fluctuations in air pressure through the auditory mechanisms and subsequently parsing the signal into higher-level linguistic units [7]. Any communication model is just a conceptual framework that simplifies our understanding of the complex communication process. This section discusses a speech transmission model, underpinned by Information Theory, known as the Speech Communication Chain. The three elements of the speech chain are production, transmission, and perception which depends upon the function of the cognitive, linguistic, physiological, and acoustic levels for both the speaker and recipient listener who are explained. Cognitive neuropsychological models that attempt to explain human communication, and language in particular, in terms of human cognition and the anatomy of the brain [8].

Figure 2.3 shows the communication chain underpinned by Shannon and Weaver's Information Theory and highlights the linguistic, physiological, and acoustic mechanisms by which humans encode, transmit, and decode meanings. This model is an extension of the Speech Chain, originally outlined by Denes and Pinson (1973) [8], as shown in Fig. 2.4 [8].

There are three main links in the communication chain:

1. Production is the process by which a human expresses himself or herself through first deciding what message is to be communicated (cognitive level). Next a plan is prepared and encoded with appropriate linguistic utterance to represent the concept (linguistic level) and, finally, produce this utterance through the suitable coordination of the vocal apparatus (physiological level) [8]. Production of a verbal utterance, therefore, takes place at three levels (Fig. 2.4):
   a. **Cognitive**: When two people talk together, the speaker sends messages to the listener in the form of words. The speaker has first to decide what he or she

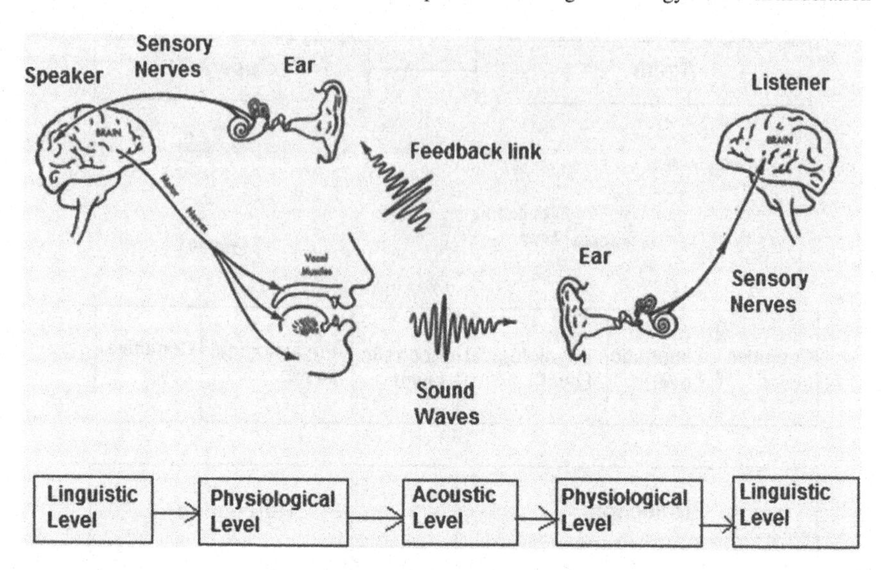

Fig. 2.4 The speech chain [8]

wants to say and then to choose the right words to put together in order to send the message. These decisions are taken in a higher level of the brain known as the cortex [8].

b. **Linguistic**: According to connectionist model, there are four layers of processing at the linguistic level: semantic, syntactic, morphological, and phonological. These work in parallel and in series, with activation at each level. Interference and misactivation can occur at any of these stages. Production begins with concepts and continues down from there [9]. Human has a bank of words stored in brains known as lexicon. It is built up over time, and the items we store are different from person to person. The store is largely dependent upon what one have been exposed to, such as the job of work it does, where one have lived, and so on. Whenever human needs to encode a word, a search is made within this lexicon in order to determine whether or not it already contains a word for the idea which is to be conveyed [8].
c. **Physiological**: Once the linguistic encoding has taken place, the brain sends small electrical signals along nerves from the cortex to the mouth, tongue, lips, vocal folds (vocal cords), and the muscles which control breathing to enable us to articulate the word or words needed to communicate our thoughts. This production of the sound sequence occurs at what is known as the physiological level, and it involves rapid, coordinated, sequential movements of the vocal apparatus.

It is clear that proper production depends upon normal perception. The prelingually deaf are clearly at a disadvantage in the learning of speech. Since normal hearers are both transmitters and perceivers of speech, they constantly perceive

their own speech as they utter it, and on the basis of this auditory feedback, they instantly correct any slips or errors. Infants learn speech articulation by constantly repeating their articulations and listening to the sounds produced [8]. This continuous feedback loop eventually results in the perfection of the articulation process [10].
2. Transmission is the sending of the linguistic utterance through some medium to the recipient. As we are only concerned with oral verbal communication here, there is only one medium of consequence, and that is air, i.e., the spoken utterance travels through the medium of air to the recipient's ear. There are, of course, other media through which a message could be transmitted. For example, written messages may be transmitted with ink and paper. However, because here, the only concern is the transmission of messages that use a so-called vocal-auditory channel, then transmission is said to occur at the acoustic level [8].
3. Reception is the process by which the recipient of a verbal utterance detects the utterance through the sense of hearing at physiological level and then decodes the linguistic expression at linguistic level. Then, the recipient infers what is meant by the linguistic expression at cognitive level [8]. Like the mirror image of production, reception also operates at the same three levels:
   a. **Physiological**: When the speaker's utterance transmitted acoustically as a speech sound wave arrives at the listener's ear, it causes his or her eardrum to vibrate. This, in turn, causes the movement of three small bones within the middle ear. Their function is to amplify the vibration of the sound wave. One of these bones, the stapes, is connected to a membrane in the wall of nerve bundle called the cochlea. The cochlea is designed to convert the vibrations into electrical signals. These are subsequently transmitted along the 30,000 or so fibres that constitute the auditory nerve to the brain of the listener. Again, this takes place at the physiological level.
   b. **Linguistic**: The listener subsequently decodes these electrical impulses in the cortex and reforms them into the word or words of the message, again at the linguistic level. The listener compares the decoded words with other words in its own lexicon. The listener is then able to determine that the word is a proper word. In order for a recipient to decode longer utterances and to interpret them as meaningful, it must also make use of the grammatical rules stored in the brain. These allow the recipient to decode an utterance such as "I have just seen a cat" as indicating that the event took place in the recent past, as opposed to an utterance such as "I will be seeing a cat" which grammatically indicates that the event has not yet happened but will happen some time in the future. Consequently, as with the agent, the recipient also needs access to a lexicon and a set of grammatical rules in order to comprehend verbal utterances.
   c. **Cognitive**: In this level, the listener must infer the speaker's meaning. A linguistic utterance can never convey all of a speaker's intended meaning. A linguistic utterance is a sketch, and the listener must fill in the sketch by inferring the meaning from such things as body language, shared knowledge, tone of voice, and so on. Humans must, therefore, be able to infer meanings in order to communicate fully.

2.3 Mechanism of Speech Perception

Speech perception is the process by which the sounds of language are heard, interpreted, and understood. The study of speech perception is closely linked to the fields of phonetics and phonology in linguistics and cognitive psychology and perception in psychology. Generally speaking, speech perception proceeds through a series of stages in which acoustic cues are extracted and stored in sensory memory and then mapped onto linguistic information. When air from the lungs is pushed into the larynx across the vocal cords and into the mouth nose, different types of sounds are produced. The different qualities of the sounds are represented in formants, which can be pictured on a graph that has time on the x-axis and the pressure under which the air is pushed, on the y-axis. Perception of the sound will vary as the frequency with which the air vibrates across time varies. Because vocal tracts vary somewhat between people, one person's vocal cords may be shorter than another's, or the roof of someone's mouth may be higher than another's, and the end result is that there are individual differences in how various sounds are produced. The acoustic signal is in itself a very complex signal, possessing extreme interspeaker and intraspeaker variability even when the sounds being compared are finally recognized by the listener as the same phoneme and are found in the same phonemic environment. Furthermore, a phoneme's realization varies dramatically as its phonemic environment varies. Speech is a continuous unsegmented sequence and yet each phoneme appears to be perceived as a discrete segmented entity. A single phoneme in a constant phonemic environment may vary in the cues present in the acoustic signal from one sample to another. Also, one person's utterance of one phoneme may coincide with another person's utterance of another phoneme, and yet both are correctly perceived [10].

2.3.1 Physical Mechanism of Perception

The primary organ responsible for initiating the process of perception is the ear. The operation of the ear has two parts—the behavior of the mechanical apparatus and the neurological processing of the information acquired. Hearing, one of the five senses of human, is the ability to perceive sound by detecting vibrations via the organ ear. Mechanical process of the ear is the beginning of the speech perception process.

The mechanism of hearing comprises a highly specialized set of structures. The anatomy of ear structures and the physiology of hearing allow sound to move from the environment through the structures of the ear and to the brain. We rely on this series of structures to transmit information, so it can be processed to get information about the sound wave. The physical structure of the ear has three sections—the outer, middle ear, and the inner ear [11]. Figure 2.5 shows the parts of the ear. The outer ear consists of the lobe and ear canal, serve to protect the more delicate parts inside. The function of the outer ear is to trap and concentrate sound waves, so their specific messages can be communicated to the other ear structures. The middle ear begins to

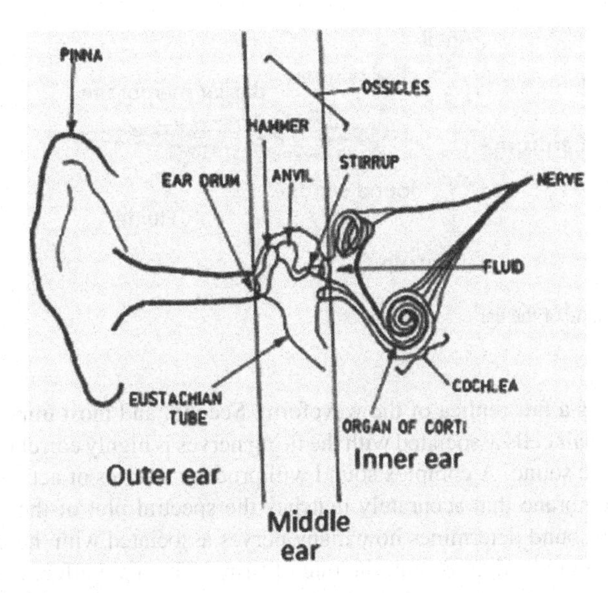

Fig. 2.5 Parts of the ear

process sounds specifically and reacts specifically to the types of sounds being heard. The outer boundary of the middle ear is the eardrum, a thin membrane that vibrates when it is struck by sound waves. The motion of the eardrum is transferred across the middle ear via three small bones named the hammer, anvil, and stirrup. These bones form a chain, which transmits sound energy from the eardrum to the cochlea. These bones are supported by muscles which normally allow free motion but can tighten up and inhibit the bones action when the sound gets too loud. The inner ear is made up of a series of bony structures, which house sensitive, specialized sound receptors. These bony structures are the cochlea, which resembles a snail shell, and the semicircular canals. The cochlea is filled with fluid which helps to transmit sound and is divided in two the long way by the basilar membrane. The cochlea contains microscopic hairs called cilia. When moved by sound waves traveling through the fluid, they facilitate nerve impulses along the auditory nerve. The boundary of the inner ear is the oval window, another thin membrane that is almost totally covered by the end of the stirrup. Sound introduced into the cochlea via the oval window flexes the basilar membrane and sets up traveling waves along its length. The taper of the membrane is such that these traveling waves are not of even amplitude the entire distance, but grow in amplitude to a certain point and then quickly fade out. The point of maximum amplitude depends on the frequency of the sound wave. The basilar membrane is covered with tiny hairs, and each hair follicle is connected to a bundle of nerves. Motion of the basilar membrane bends the hairs which in turn excites the associated nerve fibers. These fibers carry the sound information to the brain, which has two components. First, even though a single nerve cell cannot react fast enough to follow audio frequencies, enough cells are involved that the aggregate of all the

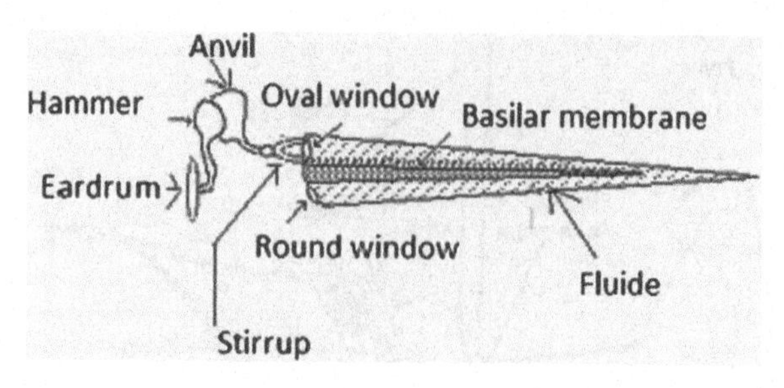

Fig. 2.6 Schematic of the ear

firing patterns is a fair replica of the waveform. Second, and most importantly, the location of the hair cells associated with the firing nerves is highly correlated with the frequency of the sound. A complex sound will produce a series of active loci along the basilar membrane that accurately matches the spectral plot of the sound. The amplitude of a sound determines how many nerves associated with the appropriate location fire, and to a slight extent, the rate of firing. The main effect is that a loud sound excites nerves along a fairly wide region of the basilar membrane, whereas a soft one excites only a few nerves at each locus [11, 12]. The schematic of the ear is shown in Fig. 2.6.

2.3.2 Perception of Sound

In the complex process of speech perception, a series of stages are involved in which acoustic cues are extracted and stored in sensory memory and then mapped onto linguistic information. Yet perception basically involves the neural operation of hearing. When sound of a particular waveform and frequency sets up a characteristic pattern of active locations on the basilar membranes, brain deals with these patterns in order to decide whether it is speech or non-speech sound and recognize that particular pattern. The brains response may be quite similar with the way brain deals with visual patterns on the retina. If a pattern is repeated enough brain learns to recognize that pattern as belonging to a certain sound, as like as human learn a particular visual pattern belonging to a certain face. The absolute position of the pattern is not very important, it is the pattern itself that is learned. Human brain possess an ability to interpret the location of the pattern to some degree, but that ability is quite variable from one person to the next [11].

Although the perception process is not so simple as described above. The process involves lots of complexities and variations. Three main effects that are crucial for understanding how the acoustic speech signal is transformed into phonetic segments and mapped into linguistic patterns are—invariance and linearity, constancy, and perceptual units. All these are described below:

- **Variance and non-linearity**: The problems of linearity and invariance are the effects of coarticulation. Coarticulation in speech production is a phenomenon in which the articulator movements for a given speech sound vary systematically with the surrounding sounds and their associated movements [13]. Simply it is the influence of articulation of one phoneme on that of another phoneme. The acoustic feature of one sound will spread themselves across those of another sound, which will lead to the problem of nonlinearity, that is, for a particular phoneme in the speech, there should be single corresponding section in the speech waveform if speech were produced linearly. However, the speech is not linear. Therefore, it is difficult to decide where one phoneme ends and the other phoneme starts. For example, it is difficult to find the location of 'th' sound and the 'uh' sound in the word 'the.' Again invariance refers to a particular phoneme having one and only one waveform representation. For example, the phoneme /i/ (the "ee" sound in "me") should have the identical amplitude and frequency as the same phoneme in "money." But this is not the case, the two waveform differ. The plosives or stop consonants, $/b/$, $/d/$, $/g/$, $/k/$, provide particular problems for the invariance assumption.
- **Inconstancy**: The perceptual system must cope with some additional sources of variability. When different talkers produce the same intended speech sound, the acoustic characteristics of their productions will differ. Women's voices tend to be higher pitched than men's, and even within a particular sex, there is a wide range of acoustic variability. Individual differences in talker's voice are due to the size of the vocal tracts, which vary from person to person. This affects the fundamental frequency of a voice. Factors like flexibility of the tongue, the state of one's vocal folds, missing teeth, etc., also will affect the acoustic characteristics of speech. In many cases, production of the same word is different by a single talker. Again in some cases, production of a particular word by one talker is acoustically similar with the production of a different word by another talker. Changes in speaking rate also affected the speech signal. Differences in voice and voiceless stops tend to decrease as speaking rate speeds. Furthermore, the nature of the acoustic changes is not predictable a priori.
- **Perceptual unit**: Phoneme and the syllable both have often been proposed as the 'primary unit of speech perception' [14, 15]. Phonemes are the unit of speech smaller than the words or syllables. Many early studies reveal that perception depends on the phonetic segmentation considering phonemes as the basic unit of speech. Phonemes contains the distinctive features that are combined into the one concurrent bundle called the phonetic segments. But it is not possible to separate formants of consonants and vowel syllable such as $/di/$. In contrast, many researchers argued that syllables are the basic unit of speech since listeners detect syllables faster than the phoneme targets [14].

2.3.3 Basic Unit of Speech Perception

The process of perceiving speech begins at the level of the sound signal and the process of audition. After processing the initial auditory signal, speech sounds are further processed to extract acoustic cues and phonetic information. This speech information can then be used for higher-level language processes, such as word recognition. The speech sound signal contains a number of acoustic cues that are used in speech perception [16]. The cues differentiate speech sounds belonging to different phonetic categories. One of the most studied cues in speech is voice onset time (VOT). VOT is a primary cue signaling the difference between voiced and voiceless stop consonants, such as 'b' and 'p'. Other cues differentiate sounds that are produced at different places of articulation or manners of articulation. The speech system must also combine these cues to determine the category of a specific speech sound. This is often thought of in terms of abstract representations of phonemes. These representations can then be combined for use in word recognition and other language processes. If a specific aspect of the acoustic waveform indicated one linguistic unit, a series of tests using speech synthesizers would be sufficient to determine such a cue or cues [17]. However, there are two significant obstacles, that is,

1. One acoustic aspect of the speech signal may cue different linguistically relevant dimensions.
2. One linguistic unit can be cued by several acoustic properties.

Speech perception is best understood in terms of an acoustic signal, phonemes, lexicals, syntax, and semantics. The human ear receives sound waves called the acoustic signal. Phonemes are the basic units of human speech, smaller than words or syllables, which allow us to distinguish between the words. Lexical relates to the words or vocabulary of a language, and syntax refers to the combination of words (or grammar) forming language. Finally, semantics refers to the meaning of the spoken message. Experts agree that speech perception does not involve simply the reception of acoustic signals and its direct translation to a semantic understanding of the message. By applying the principle of recursive decomposition, analysis of speech perception can begin with the observation that sentences are composed of words composed of syllables that are arranged grammatically. But neither the syllable nor the phoneme is the basic unit of speech perception [18].

A single phoneme is pronounced differently based on its combination with other phonemes. Consider the use of the phoneme /r/ in the words red and run. If one were to record the sound of both spoken words and then isolate the /r/ phoneme, the vowel phonemes would be distinguishable. This effect is known as coarticulation which is used to refer to the phenomenon in which the mouth position for individual phonemes in the utterance of continuous sounds incorporates the effects of the mouth position for phonemes uttered immediately before and after [19]. The articulatory gestures required to produce these words blend the phonemes, making it impossible

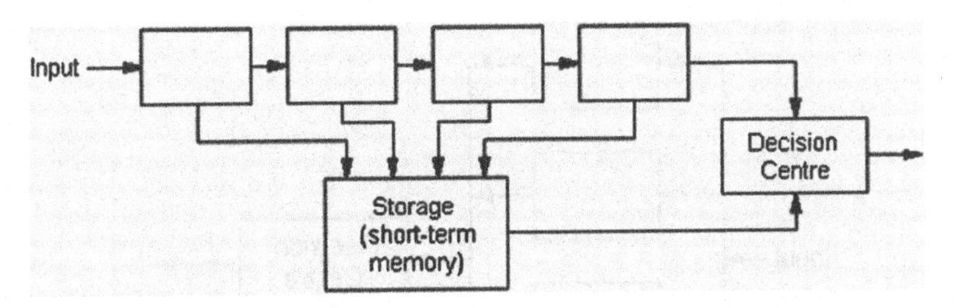

Fig. 2.7 Series processing

to distinguish them as separate units. Thus, the basic unit of speech perception is coarticulated phonemes [18].

2.4 Theories and Models of Spoken Word Recognition

The basic theories of speech perception provide a better explanation about how the process works. In series models, there is a sequential order to each subprocess. A contrast to series models is parallel models which feature several subprocesses acting simultaneously [18].

The perception of speech involves the recognition of patterns in the acoustic signal in both time and frequency domains. Such patterns are realized acoustically as changes in amplitude at each frequency over a period of time [10]. Most theories of pattern processing involve series, arrays, or networks of binary decisions. In other words, at each step in the recognition process, a yes/no decision is made as to whether the signal conforms to one of two perceptual categories. The decision thus made usually affects which of the following steps will be made in a series of decisions. Figure 2.7 shows the series processing scheme [10]. The series model of Bondarko and Christovich begins with the listener receiving speech input and then subjecting it to auditory analysis. Phonetic analysis then passes the signal to a morphological analysis block. Finally, a syntactic analysis of the signal results in a semantic analysis of the message [18]. Each block in the series is said to reduce/refine the signal and pass additional parameters to the next level for further processing. Liberman proposes a top–down, bottom–up series model which demonstrates both signal reception and comprehension with signal generation and transmission in the following format: acoustic structure sound, speech phonetic structure, phonology, surface structure, syntax, deep structure, semantics, and conceptual representation. Series models imply that decisions made at one level of processing affect the next block, but do not receive feedback. Lobacz suggests that speech perception is, in reality, more dynamic and that series models are not an adequate representation of the process [18].

The decision made in series processing usually affects which following steps will be made in a series of decisions. If the decision steps are all part of a serial processing

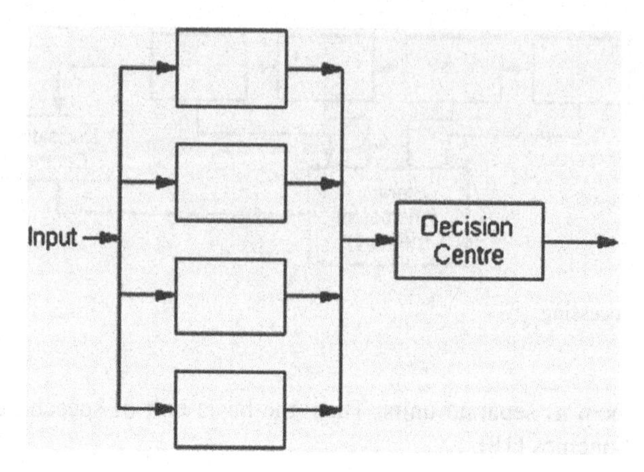

Fig. 2.8 Parallel processing

chain, then a wrong decision at an early stage in the pattern recognition process may cause the wrong questions to be asked in subsequent steps. It means that each step may be under the influence of previous steps. A serial processing system also requires a facility to store each decision so that all the decisions can be passed to the decision center when all the steps have been completed. Because of all these problems with serial processing strategies, most speech perception theorists prefer at least some sort of parallel processing as shown in Fig. 2.8. In parallel processing, all questions are asked simultaneously, that is, all cues or features are examined at the same time, and so processing time is very short no matter how many features are examined. Since all tests are processed at the same time, there is no need for the short-term memory facility, and further, there is also no effect of early steps on following steps, that is, no step is under the influence of a preceding step [10]. Thus, it can be said that the need to extract several phonemes from any syllable is taken as evidence that speech perception is a parallel process. Parallel models demonstrate simultaneous activity in the subprocesses of speech perception, showing that a process at one level may induce processes at any other level, without hierarchical information transmission. Lobacz cites Fants model as an example. The model consists of five successive stages of processing:

- Acoustic parameter extraction,
- Microsegment detection,
- Identification of phonetic elements (phonetic transcription),
- Identification of sentence structure (identification of words), and
- Semantic interpretation

Each stage includes stores of previously acquired knowledge. The stores hold inventories of units extracted at the given level with constraints on their occurrence in messages. The system also includes comparators that are placed between the successive stages of processing. The model provides for a possibility of direct connection between the lowest and the highest levels [18].

Further, speech perception theories can be considered to be of two types or a combination of both:

1. **Passive or Non-mediated theories**: These theories are based on the assumption that there is some sort of direct relationship between the acoustic signal and the perceived phoneme. In other words, perceptual constancy is in some way matched to a real acoustic constancy. These theories tend to concentrate on discovering the identity of such constant perceptual cues, and on the way, the auditory system might extract them from the acoustic signal. In one way or another, these theories are basically filtering theories and do not involve the mediation of higher cognitive processes in the extraction of these cues. These higher processes are restricted to making a decision based on the features or cues which have been detected or extracted closer to the periphery of the auditory system [10].
2. **Active or mediated theories**: These theories, on the other hand, suggest that there is no direct relationship between the acoustic signal and the perceived phoneme but rather that some higher-level mediation is involved in which the input pattern is compared with an internally generated pattern [10].

Spoken word recognition is a distinct subsystem providing the interface between low-level perception and cognitive processes of retrieval, parsing, and interpretation of speech. The process of recognizing a spoken word starts from a string of phonemes, establishes how these phonemes should be grouped to form words, and passes these words onto the next level of processing.

Research on the discrimination and categorization of phonetic segments was the key focus of the works on speech perception before 1970s. The processes and representations responsible for the perception of spoken words became a primary object of scientific inquiry with a curiosity of disclosing the cause and methods of how the listener perceives fluent speech. The following sections provides a brief description of some renown models and theory of spoken word recognition found in the literature.

2.4.1 Motor Theory of Speech Perception

Beginning in the early 1950s, Alvin Liberman, Franklin Cooper, Pierre Delattre, and other researchers at the Haskins Laboratories carried out a series of landmark studies on the perception of synthetic speech sounds. This work provided the foundation of what is known about acoustic cues for linguistic units such as phonemes and features and revealed that the mapping between speech signals and linguistic units is quite complex. In time, Liberman and his colleagues became convinced that perceived phonemes and features have a simpler (i.e., more nearly one-to-one) relationship to articulation than to acoustics, and this gave rise to the motor theory of speech perception. Every version has claimed that the objects of speech perception are articulatory events rather than acoustic or auditory events [20].

What a listener does, according to this theory, is to refer the incoming signal back to the articulatory instructions that the listener would give to the articulators in order to produce the same sequence. The motor theory argues that the level of articulatory or motor commands is analogous to the perceptual process of phoneme perception and that a large part of both the descending pathway (phoneme to articulation) and the ascending pathway (acoustic signal to phoneme identification) overlaps. The two processes represent merely two-way traffic on the same neural pathways. The motor theory points out that there is a great deal of variance in the acoustic signal and that the most peripheral step in the speech chain which possesses a high degree of invariance is at the level of the motor commands to the articulatory organs. The encoding of this invariant linguistic information is articulatory and so the decoding process in the auditory system must at least be analogous. The motor theorists propose that 'there exists an overlapping activity of several neural networks—those that supply control signals to the articulators, and those that process incoming neural patterns from the ear and that information can be correlated by these networks and passed through them in either direction' [10].

This theory suggests that there exists a special speech code (or set of rules) which is specific to speech and which bridges the gap between the acoustic data and the highly abstract higher linguistic levels. Such a speech code is unnecessary in passive theories, where each speech segment would need to be represented by a discrete pattern, that is, a template somehow coded into the auditory system at some point. The advantage of a speech code or rule set is that there is no need for a vast storage of templates since the input signal is converted into a linguistic entity using those rules. The rules achieve their task by a drastic restructuring of the input signal. The acoustic signal does not in itself contain phonemes which can be extracted from the speech signal (as suggested by passive theories), but rather contains cues or features which can be used in conjunction with the rules to permit the recovery of the phoneme which last existed as a phonemic entity at some point in the neuromuscular events which led to its articulation. This, it is argued, is made evident by the fact the speech can be processed 10 times faster than can non-speech sounds. Speech perception is seen as a distinct form of perception quite separate from that of non-speech sound perception. Speech is perceived by means of categorical boundaries, while non-speech sounds are tracked continuously. Like the proponents of the neurological theories proposed by Abbs and Sussman in 1971 [10], the motor theorists believe that speech is perceived by means of a special mode, but they believe that this is not based directly on the recognition of phonemes embedded in the acoustic signal but rather on the gating of phonetically processed signals into specialized neural units. Before this gating, both phonetic and non-phonetic processings have been performed in parallel, and the non-phonetic processing is abandoned when the signal is identified as speech [10].

Speech is received serially by the ear, and yet must be processed in some sort of parallel fashion since not all cues for a single phoneme coexist at the same point in time, and the boundaries of the cues do not correspond to any phonemic boundaries. A voiced stop requires several cues to enable the listener to distinguish it. VOT, that is, voice onset time, is examined to enable a voiced or voiceless decision, but this is essentially a temporal cue and can only be measured relative to the position of the release burst. The occlusion itself is a necessary cue for stops in general in a

Vowel–Consonant–Vowel (VCV) environment, and yet it does not coexist in time with any other cue. The burst is another general cue for stops, while the following aspiration (if present) contains a certain amount of acoustic information to help in the identification of the stop's place of articulation and further helps in the identification of positive VOT and thus of voiceless stops. The main cues to the stop's place of articulation are the formant transitions into the vowel, and these in no way co-occur with the remainder of the stop's cues. This folding over of information onto adjacent segments, which we know as coarticulation, far from making the process of speech perception more confusing, actually helps in the determination of the temporal order of the individual speech segments, as it permits the parallel transmission via the acoustic signal of more than one segment at a time [10].

2.4.2 Analysis-by-Synthesis Model

Stevens and Halle in 1967 have postulated that 'the perception of speech involves the internal synthesis of patterns according to certain rules, and a matching of these internally generated patterns against the pattern under analysis...moreover, ...the generative rules in the perception of speech [are] in large measure identical to those utilized in speech production, and that fundamental to both processes [is] an abstract representation of the speech event.' In this model, the incoming acoustic signal is subjected to an initial analysis at the periphery of the auditory system. This information is then passed upward to a master control unit and is there processed along with certain contextual constraints derived from preceding segments. This produces an hypothesized abstract representation defined in terms of a set of generative rules. This is then used to generate motor commands, but during speech perception, articulation is inhibited and instead the commands produce a hypothetical auditory pattern which is then passed to a comparator module. It compares this with the original signal which is held in a temporary store. If a mismatch occurs, the procedure is repeated until a suitable match is found [10]. Figure 2.9 shows the Analysis-by-Synthesis model.

2.4.3 Direct Realist Theory of Speech Perception

Starting in the 1980s, an alternative to Motor Theory, referred to as the Direct Realist Theory (DRT) of speech perception, was developed by Carol Fowler, also working at the Haskins Laboratories. Like Motor Theory, DRT claims that the objects of speech perception are articulatory rather than acoustic events. However, unlike Motor Theory, DRT asserts that the articulatory objects of perception are actual, phonetically structured, vocal tract movements, or gestures, and not events that are causally antecedent to these movements, such as neuromotor commands or intended gestures. DRT also contrasts sharply with Motor Theory in denying that specialized, that is,

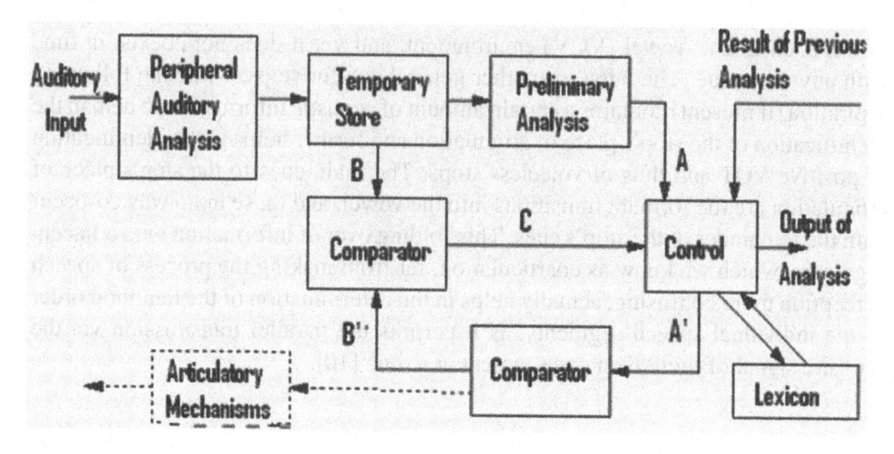

Fig. 2.9 Analysis-by-synthesis model (after Stevens 1972)

speech-specific or human-specific, mechanisms play a role in speech perception. This theory is elegantly summarized by Fowler in 1996 in the following passage: Perceptual systems have a universal function. They constitute the sole means by which animals can know their niches. Moreover, they appear to serve this function in one way. They use structure in the media that has been lawfully caused by events in the environment as information for the events. Even though it is the structure in media (light for vision, skin for touch, air for hearing) that sense organs transduce, it is not the structure in those media that animals perceive. Rather, essentially for their survival, they perceive the components of their niche that caused the structure [20]. Thus, according to DRT, a talker gestures (e.g., the closing and opening of the lips during the production of */pa/*) structure the acoustic signal, which then serves as the informational medium for the listener to recover the gestures. The term direct in direct realism is meant to imply that perception is not mediated by processes of inference or hypothesis testing, rather, the information in the acoustic signal is assumed to be rich enough to specify (i.e., determine uniquely) the gestures that structure the signal. To perceive the gestures, it is sufficient for the listener simply to detect the relevant information. The term realism is intended to mean that perceivers recover actual (physical) properties of their niche, including, in the case of speech perception, phonetic segments that are realized as sets of physical gestures. This realist perspective contrasts with a mentalistic view that phonetic segments are internally generated, the creature of some kind of perceptual-cognitive process [20].

2.4.4 Cohort Theory

The original Cohort theory was proposed by Marslen-Wilson and Tyler in 1980 [21]. According to Cohort theory, various language sources like lexical, syntactic,

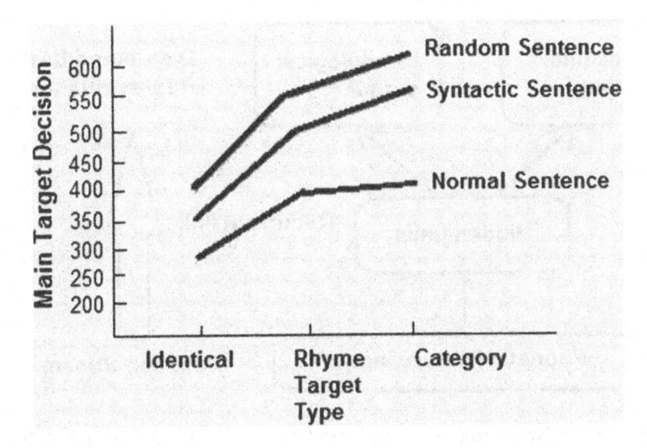

Fig. 2.10 Detection time in word sentence presented in sentences in the Cohort theory (Marslen-Wilson and Tyler 1980)

semantic, etc., interact with each other in complex ways to produce an efficient analysis of spoken language. It suggests that input in the form of a spoken word activates a set of similar items in the memory, which is called word initial cohort. The word initial cohort consisted of all words known to the listener that begin with the initial segment or segments of the input words. For example, the word elephant may activate in the word initial cohort members echo, enemy, elder, elevator, and so on. The processing of the words started with both bottom–up and top–down information and continuously eliminated those words if they are found to be not matching with the presented word. Finally, a single word remains from the word initial cohort. This is the 'recognition point' of the speech sound [14, 21].

Marslen-Wilson and Tyler performed a word-monitoring task, in which participants had to identify prespecified target words presented in spoken sentence as rapidly as possible. There were normal sentences, syntactic sentences, that is, grammatically correct but meaningless and random sentences, that is, unrelated words. The target was a member of given category, a word that rhymed with a given word, or a word identical to a given word. Figure 2.10 shows the graph of detection time versus target. According to the theory, sensory information from the target word and contextual information from the rest of the sentence is both used at the same time [21].

In 1994, Marslen-Wilson and Warren revised the Cohort theory. In the original version, words were either in or out of the word cohort. But in the revised version, words candidate vary in the activation level, and so membership of the word initial cohort plays an important role in the recognition. They suggested that word initial cohort contains words having similar initial phoneme [21].

The latest version of the cohort is the distributed Cohort model proposed by Gaskell and Marslen-Wilson in 1997, where the model has been rendered in a connectionist architecture as shown in Fig. 2.11. This model differs from the others in the fact that it does not have lexical representations. Rather, a hidden layer mediates between the phonetic representation on the one side and phonological and semantic

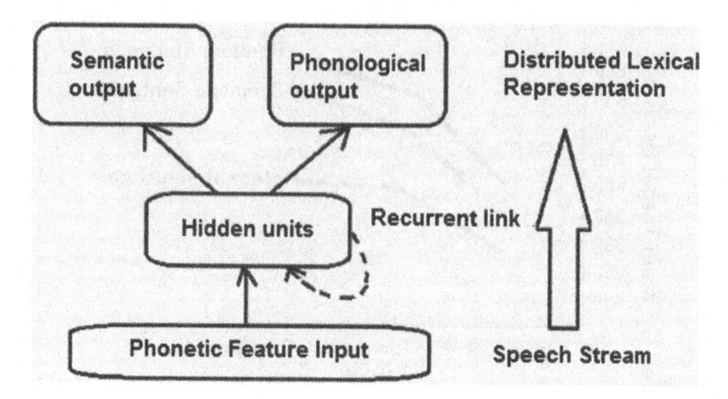

Fig. 2.11 The connectionist distributed Cohort model of Gaskell and Marslen-Wilson (1997)

representations on the other side. The representations at each level are distributed. In Cohort, an ambiguous phonetic input will yield interference at the semantic and phonological levels. Input to the recurrent network is passed sequentially, so given the beginning of a word, the output layers will have activation over the various lexical possibilities, which as the word progresses toward its uniqueness point, will subside [22].

In this model, lexical units are points in a multidimensional space, represented by vectors of phonological and semantic output nodes. The phonological nodes contain information about the phonemes in a word, whereas the semantic nodes contain information about the meaning of the words. The speech input maps directly and continuously onto this lexical knowledge. As more bottom–up information comes in, the network moves toward a point in lexical space corresponding to the presented word. Activation of a word candidate is thus inversely related to the distance between the output of the network and the word representation in lexical space. A constraining sentence context functions as a bias: The network shifts through the lexical space in the direction of the lexical hypotheses that fit the context. However, there is little advantage of a contextually appropriate word over its competitors early on in the processing of a word. Only later, when a small number of candidates still fits the sensory input, context starts to affect the activation levels of the remaining candidates more significantly. There is also a mechanism of bottom–up inhibition, which means that in case the incoming sensory information no longer fits that of the candidate, the effects of the sentence context are overridden [14].

2.4.5 Trace Model

The TRACE model of spoken word recognition, proposed by McClelland and Elman, in 1986, is an interactive activation network model . According to this, model speech perception can be described as a process, in which speech units are arranged into

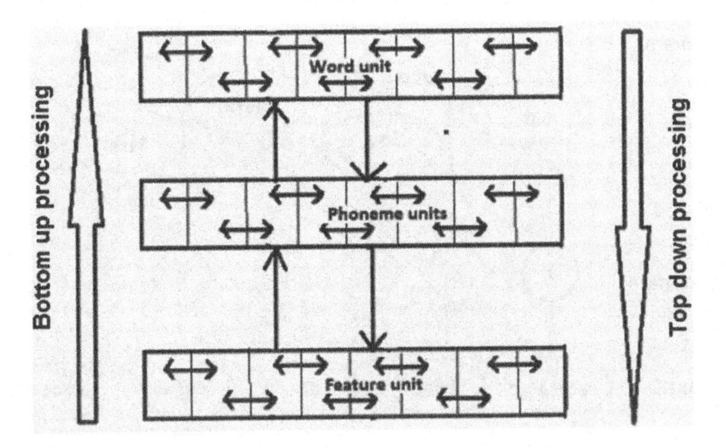

Fig. 2.12 Basic Trace model

levels which interact with each other. There are three levels: features, phonemes, and words. There are individual processing units or nodes at three different levels. For example, within the feature level, there are individual nodes that detect voicing. Feature nodes are connected to phoneme nodes, and phoneme nodes are connected to word nodes. Connection between levels operate in both directions. The processing units or nodes share excitatory connection between levels and share inhibitory links within levels [14, 23]. For example, to perceive a /k/ in 'cake,' the /k/ phoneme and corresponding featural units share excitatory connections. Again, /k/ would have an inhibitory connection with the vowel sound in 'cake' /eI/.

Each processing units corresponds to the possible occurrence of a particular linguistic unit (feature, phoneme, or word) at a particular time within a spoken input. Activation of a processing unit reflects the state of combined evidence within the system for the presence of that linguistic unit. When input is presented at the feature layer, it is propagated to the phoneme layer and then to the lexical layer. Processing proceeds incrementally with between-layer excitation and within-layer competitive inhibition. Excitation flow is bidirectional, that is, both bottom–up and top–down processing interact during perception. Bottom–up activation proceeds upward from the feature level to the phoneme and phoneme level to the word level. Whereas top–down activation proceeds in the opposite from the word level to the phoneme level and phoneme level to the feature level as shown in Fig. 2.12. As excitation and inhibition spread among nodes, a pattern of activation develops. The word that is recognized is determined by the activation level of the possible candidate words [23].

Trace is a model of accessing the lexicon with interlevel feedback. Feedback in the Trace model improves segmental identification under imperfect hearing conditions. In interactive activation networks, such as Trace, each word or lexical candidate is represented by a single node, which is assigned an activation value. The activation of a lexical node increases as the node receives more perceptual input and decreases when subject to inhibition from other words [21].

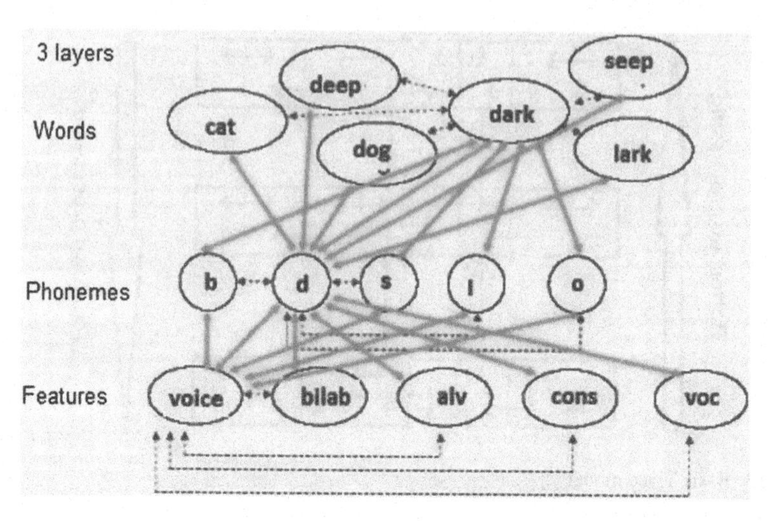

Fig. 2.13 Interaction in Trace model

In Trace model, activations are passed between levels, and thus, it confirms the evidence of a candidate to a given feature, phoneme, or words. For example, if the voicing speech is presented, that will activate the voiced feature at the lowest level of the model, which in turn pass its activation to all voiced phoneme in the next level of phonemes, and this activates the words containing those phonemes in the word level. The important fact is that through lateral inhibition among units, one candidate in a particular level may activate some similar units, in order to help the best candidate word to win the recognition. Such as, the word 'cat' at the lexical level will send inhibitory information to the word 'pat' in the lexical unit. Thus, the output may be one or more suggested candidate. Figure 2.13 shows the interaction in a trace model through inhibitory and excitatory connection. Each unit connects directly with every unit in its own layer, and in the adjacent layers. Connections between layer are excitatory that activates units in other layers. Connections within a layer are inhibitory that activates units in same layer.

2.4.6 *Shortlist Model*

Dennis Norris, in 1994, have proposed the Shortlist model, which is a connectionist model of spoken word recognition. According to Norris, a Shortlist of word candidates is derived at the first stage of the model. The list consists of lexical items that match the bottom–up speech input. This abbreviated list of lexical items enters into a network of word units in the later stage, where lexical units compete with one another via lateral inhibitory links [14]. Unlike trace flow of information in the Shortlist model is unidirectional and only bottom–up processing involves. The superiority

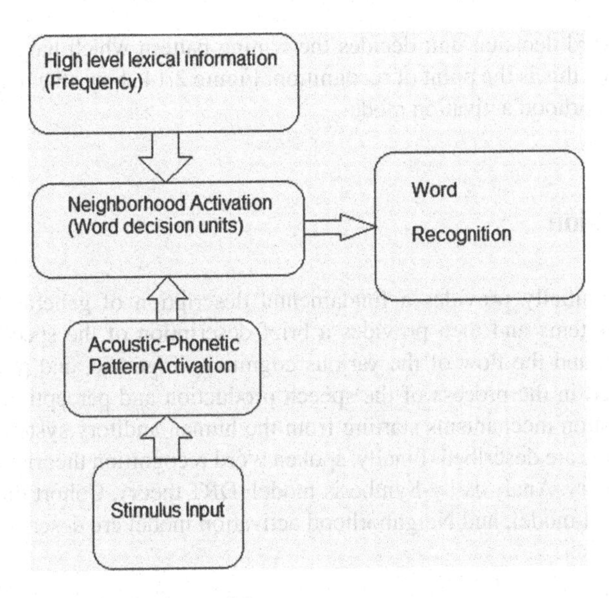

Fig. 2.14 Neighborhood activation model

of Shortlist comes from the fact that it provides an account for lexical segmentation in fluent speech.

2.4.7 Neighborhood Activation Model

According to the Neighborhood Activation Model, proposed by Luce, Pisoni and Goldinger in 1990, and revised by Luce and Pisoni in 1998, spoken word recognition is a special kind of pattern recognition task performed by human. Some similar acoustic-phonetic patterns are activated in the memory by the stimulus speech input. Once the patterns are activated, the word decision units interact with each of the patterns and tries to decide which pattern best matches the input pattern. The level of activation depends on the similarity of the pattern to the input speech. The more similar the pattern, the higher is its activation level [14, 24]. The word decision unit performs some probabilistic computation to measure the similarity of the patterns. It considers following facts:

1. The frequency of the word to which the pattern corresponds.
2. The activation level of the pattern, which depends on the patterns match to the input.
3. The activation levels and frequencies of all other words activated in the system.

Finally, the word decision unit decides the wining pattern which have the highest probability, and this is the point of recognition. Figure 2.14 shows the diagrammatic view of neighborhood activation model.

2.5 Conclusion

This chapter initially provides a fundamental description of generalized speech recognition systems and then provides a brief description of the speech communication chain and the flow of the various cognitive, linguistic, and physiological events involved in the process of the speech production and perception. Next, the speech perception mechanisms starting from the human auditory system and hearing mechanisms are described. Finally, spoken word recognition theories and model like Motor theory, Analysis-by-Synthesis model, DRT theory, Cohort theory, Trace model, Shortlist model, and Neighborhood activation model are described in a nutshell.

References

1. Automatic Speech Recognition. Available via www.kornai.com/MatLing/asr.doc
2. Speech Recognition. Available via www.learnartificialneuralnetworks.com/speechrecognition.html
3. Discrete Speech Recognition Systems. Available via http://www.gloccal.com/discrete-speech-recognition-systems.html
4. Types of Speech Recognition. Available via http://www.lumenvox.com/resources/tips/types-of-speech-recognition.aspx
5. Huang XD (1991) A study on speaker-adaptive speech recognition. In: Proceedings of the workshop on speech and natural language, pp 278–283
6. Rabiner LR, Schafer RW (2009) Digital processing of speech signals. Pearson Education, Dorling Kindersley (India) Pvt Ltd, Delhi
7. Narayan CR (2006) Acoustic-perceptual salience and developmental speech perception. Doctor of Philosophy Dissertation, University of Michigan
8. Speech Therapy-Information and Resources. Available via http://www.Speech-therapy-information-and-resources.com/communication-model.html
9. Levelt WJM (1989) Speaking, from intention to articulation. MIT Press, England
10. Mannell R (2013) Speech perception background and some classic theories. Department of Linguistics, Feculty of Human Sciences, Macquarie university, Available via http://clas.mq.edu.au/perception/speechperception/classic_models.html
11. Elsea P (1996) Hearing and perception. Available via http://artsites.ucsc.edu/EMS/Music/tech_background/TE-03/teces_03.html
12. The Hearing Mechanism. www.healthtree.com. 20 July 2010
13. Ostry DJ, Gribble PL, Gracco VL (1996) Coarticulation of jaw movements in speech production: is context sensitivity in speech kinematics centrally planned? J Neurosci 16:1570–1579
14. Jusczyk PW, Luce PA (2002) Speech perception and spoken word recognition: past and present. Ear Hear 23(1):2–40
15. Pallier C (1997) Phonemes and syllables in speech perception: size of attentional focus in French. In: Proceedings of EUROSPEECH, ISCA

16. Klatt DH (1976) Linguistic uses of segmental duration in English: acoustic and perceptual evidence. J Acoust Soc Am 59(5):1208–1221
17. Nygaard LC, Pisoni DB (1995) Speech perception: new directions in research and theory. Handbook of perception and cognition: speech, language, and communication. Elsevier, Academic Press, Amsterdam, pp 6396
18. Anderson S, Shirey E, Sosnovsky S (2010) Department of Information Science. University of Pittsburgh,Speech Perception, Downloaded on Sept 2010
19. Honda M (2003) Human speech production mechanisms. NTT CS Laboratories. Selected Papers (2):1
20. Diehl RL, Lotto AJ, Holt LL (2004) Speech perception. Annu Rev Psychol 55:149–179
21. Eysenck MW (2004) Psychology: an international perspective. Psychology Press
22. Bergen B (2006) Speech perception. Linguistics 431/631: connectionist language modeling. Available via http://www2.hawaii.edu/bergen/ling631/lecs/lec10.html
23. McClelland JL, Mirman D, Holt LL (2006) Are there interactive processes in speech perception? Trends Cogn Sci 10:8
24. Theories of speech perception, Handout 15, Phonetics Scarborough (2005) Available via http://www:stanford.edu/class/linguist205/indexfiles/Handout%2015%20%20Theories%20of%20Speech%20Perception.pdf

Chapter 3
Fundamental Considerations of ANN

Abstract Fundamental considerations of Artificial Neural Network is described in this chapter. Initially, the analogy of artificial neuron with the biological neuron is explained along with a description of the commonly used activation functions. Then, two basic ANN learning paradigms namely supervised and unsupervised learning are described. A brief note on prediction and classification using ANN is given next. Finally, primary ANN topologies like Multilayer Perceptron (MLP), Recurrent Neural Network (RNN), Probabilistic Neural Network (PNN), Learning Vector Quantization (LVQ), and Self-Organizing Map (SOM) are explained theoretically which are extensively used in the work described throughout the book.

Keywords Artificial neural network · Neuron · McCulloch-Pitts neuron · Learning · Prediction · Classification

3.1 Introduction

Artificial Neural Networks (ANNs) are nonparametric prediction tools that can be used for a host of pattern classification purposes. It is an information processing paradigm that is inspired by the way biological nervous systems, such as the brain, process information. The key element of this paradigm is the novel structure of the information processing system. It is composed of a large number of highly interconnected processing elements (neurones) working in unison to solve specific problems. ANNs, like people, learn by example. An ANN is configured for a specific application, such as pattern recognition or data classification, through a learning process. ANNs are adjusted, or trained, so that a particular input leads to a specific target output. ANNs can be trained to perform complex functions in various fields, including pattern recognition, identification, classification, speech, vision, and control systems. ANNs can also be trained to solve problems that are difficult for conventional computers or human beings [1].

M. Sarma and K. K. Sarma, *Phoneme-Based Speech Segmentation Using Hybrid Soft Computing Framework*, Studies in Computational Intelligence 550,
DOI: 10.1007/978-81-322-1862-3_3,

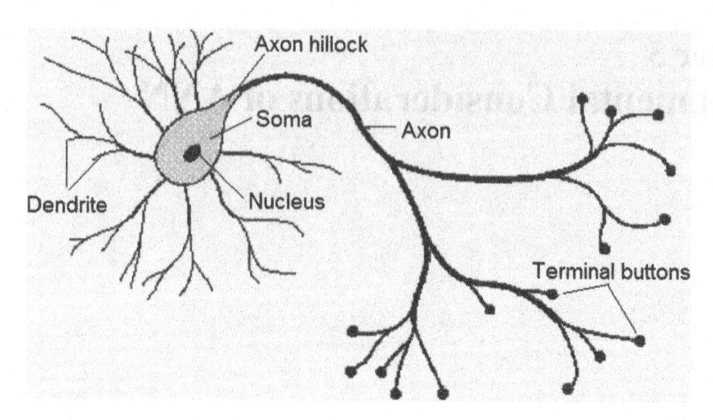

Fig. 3.1 A generic biological neuron

There is a close analogy between the structure of a biological neuron (i.e., a brain or nerve cell) and the processing element of the ANN. A biological neuron has three types of components: its *dendrites*, *soma* and *axon*. Many dendrites receive signals from other neurons. The signals are electric impulses that are transmitted across a synaptic gap by means of a chemical process. The action of the chemical transmitter modifies the incoming signal in a manner similar to the action of the weights in an ANN. The soma or call body sums the incoming signals. when sufficient input is received, the cell transmits a signal over its axon to other cells. At any instant of time, a cell may transmit or not, so that transmitted signals can be treated as binary. However, the frequency of firing varies and can be viewed as a signal of either greater or lesser magnitude. This corresponds to looking at discrete time steps and summing all activity, i.e., signals received or signals sent, at a particular point in time [2]. The transmission of the signals from a particular neuron is accomplished by an action potential resulting from differential concentrations of ions on either side of the neuron's axon sheath. The ions most directly involved are potassium, sodium, and chloride [2]. Figure 3.1 shows a generic biological neuron, together with axons from other neurons.

The fundamental information processing unit of the operation of the ANN is the McCulloch-Pitts neuron (1943). When creating a functional model of the biological neuron, there are three basic components of importance. First, the synapses of the neuron are modeled as weights. The strength of the connection between an input and a neuron is noted by the value of the weight. Negative weight values reflect inhibitory connections, while positive values designate excitatory connections. The next two components model the actual activity within the neuron cell. An adder sums up all the inputs modified by their respective weights. This activity is referred to as linear combination. Finally, an activation function controls the amplitude of the output of the neuron. An acceptable range of output is usually between 0 and 1, or −1 and 1 [3]. The block diagram of Fig. 3.2 shows the model of a neuron, which forms the basis for designing ANN. The three basic elements of the neuronal model are explained below:

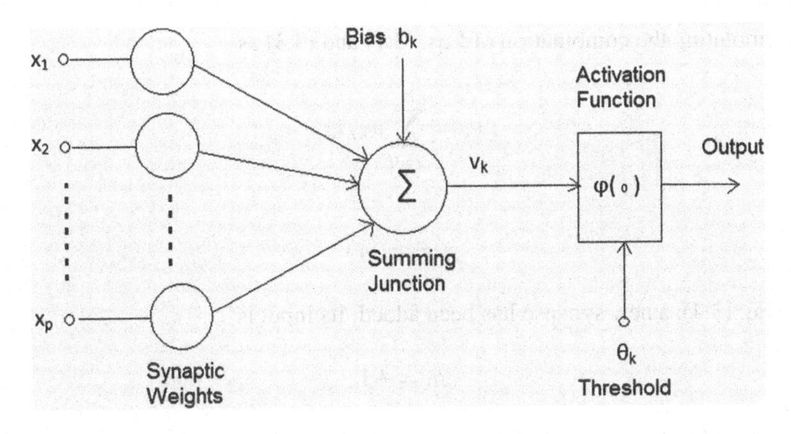

Fig. 3.2 A nonlinear model of a neuron

1. A set of synapses or connecting links, each of which is characterized by a weight or strength of its own. Specially, a signal x^j at the input of synapse j connected to neuron k is multiplied by the synaptic weight w^{kj}. The first subscript of synaptic weight w^{kj} refers to the neuron in question, and the second subscript refers to the input end of the synapse to which the weight refers. Unlike a synapse in the brain, the synaptic weight of an artificial neuron may lie in a range that includes negative as well as positive values [3].
2. An adder for summing the input signal, weighted by the respective synapses of the neuron; the operations described here constitute a linear combiner [3].
3. An activation function for limiting the amplitude of the output of a neuron. The activation function is also referred to as a squashing function in that it squashes (limits) the permissible amplitude range of the output signal to some finite value [3].

From this model, the interval activity of a neuron k can be shown to be

$$u_k = \sum_{j=1}^{m} w_{kj} x_j \tag{3.1}$$

and

$$y_k = \varphi(u_k + b_k) \tag{3.2}$$

where $x_1, x_2, \ldots, x_k$ are the input signals, $w_{k1}, w_{k2}, \ldots, w_{km}$ are the synaptic weights of neuron k, b_k is the bias, $\varphi(.)$ is the activation function, and y_k is the output signal of the neuron. The use of bias b_k has the effect of applying an *affine transformation* to the output u_k of the linear combiner in the model of Fig. 3.2, as shown by

$$\nu_k = u_k + b_k \tag{3.3}$$

Formulating the combination of Eqs. (3.1) and (3.3) as

$$\nu_k = \sum_{j=0}^{m} w_{kj} x_j \tag{3.4}$$

and

$$y_k = \varphi(\nu_k) \tag{3.5}$$

In Eq. (3.4), a new synapse has been added. Its input is

$$x_0 = +1 \tag{3.6}$$

and its weight is

$$w_{k0} = b_k \tag{3.7}$$

3.2 Learning Strategy

Learning algorithms define an architectural-dependent procedure to encode pattern information into weights to generate these internal models. Learning proceeds by modifying connection strengths [4].

In biological system, learning alters the efficiency of a synapse, both in the amount of neurotransmitter released by a synaptic terminal and in the physical structure of the axonal-dendritic junction that would allow greater or lesser amount of the neurotransmitter to effect postsynaptic channels. In artificial systems, learning changes the synaptic weights in the model [4].

Learning is data driven. The data might take the form of a set of input–output patterns derived from a possibly unknown probability distribution. Here, the output pattern might specify a desired system response for a given pattern, and the issue of learning would involve approximating the unknown function as described by the given data. Alternatively, the data might comprise pattern that naturally cluster into some number of unknown classes, and the learning problem might involve generating a suitable classification of the sample [4]. The nature of the problem of learning allows to demarcate learning algorithms into two categories:

1. Supervised Learning.
2. Unsupervised Learning.

The data available for learning may comprise of a set of discrete samples $\tau = \{(X_k, D_k)\}_{k=1}^{Q}$ drawn from the pattern space where each sample relates an input vector $X_k \in R^n$ to an output vector $D_K \in R^p$. The set of samples describe the behavior of an unknown function $f : R^n \longrightarrow R^p$ which is to be characterized using a learning system. Figure 3.3 portrays the structure of a supervised learning system. When an input X_k is presented to the system, it generates a response S_k . Supervised

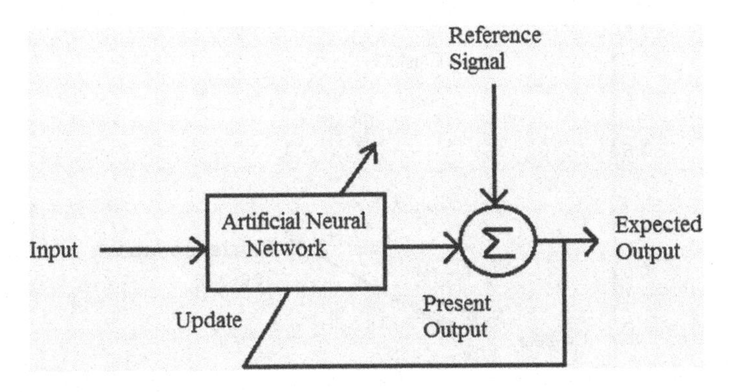

Fig. 3.3 Supervised learning procedure

learning employs the *teaching input* D_k, the associates of X_k to reduce the error (D_k, S_k) in the response the system. Input–output sample pairs are employed to train the network through a simple form of error correction learning or gradient descent weight adaptation. These procedure are based on global error measures derived from the difference between the desired (D_k) and actual (S_k) output of the network. The implementation of learning in a supervised system is usually in the form of difference equations that are designed to work with such global information. Supervised learning encodes a behavioristic pattern into the network by attempting to approximate the function that underlines the data set. The system generates an output D_k in response to an input X_k, and the system has learnt the underlying map if a stimulus X'_k close to elicits a response S'_k which is sufficiently close to D_k.

The alternate way of learning is to simply provide the system with an input X_k, and allow it to self-organize its parameters, which are the weight of the network to generate internal prototypes of sample vector. An unsupervised learning system attempts to represent the entire data set by employing a small number of prototype vectors—enough to allow the system to retain a desired level of discrimination between samples. As new samples continuously buffer the system, the prototypes will be in a state of constant flux. There is no teaching input. This kind of learning is often called *adaptive vector quantization (AVQ)* and is essentially unsupervised in nature. Figure 3.4 shows one example where the sample space is quantized by exactly two vectors which are the centroids of the data clusters.

Given a set of data samples X_i with $X_i \in R_n$, it is possible to identify well defined 'clusters', where each cluster defines a class of vectors which are similar in some broad sense. Cluster helps establish a classification structure within a data set that has no categories defined in advance. Classes are derived from clusters by appropriate labeling. The goal of pattern classification is to assign an input pattern to one of a finite number of classes. Each cluster or class can be characterized by a prototype vector which is a 'codebook' vector that best represents members of the cluster. In unsupervised systems, weights of the ANN undergo a process of self-organization—without the intervention of a desired behavioral stimulus—to create

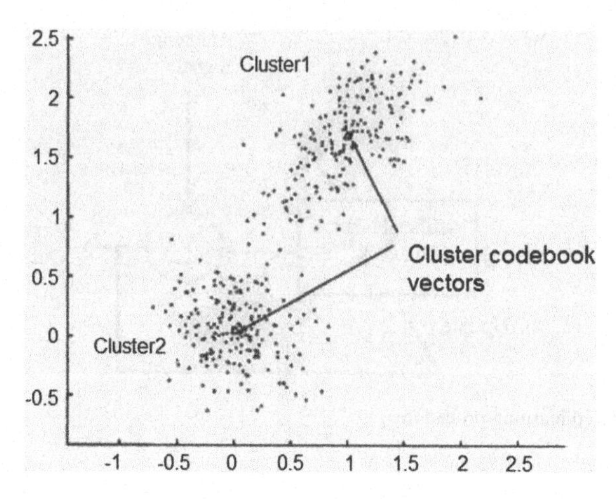

Fig. 3.4 Quantization of pattern clusters by prototype or codebook vectors

clusters of pattern represented by a codebook vectors. Learning in an unsupervised system is often driven by a competitive-cooperative process where individual neurons complete and cooperate with each other to update their weights based on the present input. Only winning neurons or clusters of neurons learn. Unsupervised learning systems embody complicated time dynamics. Learning in such systems is usually implemented in the form of differential equations that are designed to work with information available only locally to a synapse.

3.3 Prediction and Classification Using ANN

Accuracy of a prediction and estimation process depends to a large extent on a higher value of proper classification. ANNs have been one of the preferred classifiers for their ability to provide optimal solution to a wide class of arbitrary classification problems [3]. MLPs trained with (error) backpropagation (BP) in particular have found wide spread acceptance in several classification, recognition, and estimation applications. The reason is that the MLPs implement linear discriminants but in a space where inputs are mapped nonlinearly. The strength of this classification is derived from the fact that MLPs admit fairly simple algorithms where the form of nonlinearity can be learned from training data [3]. The classification process involves defining a decision rule so as to provide boundaries to separate one class of data from another. It means placing the *j*th input of the input vector $\mathbf{x(j)}$ in the proper class among M outputs of a classifier.

Classification by an ANN involves training it so as to provide an optimum decision rule for classifying all the outputs of the network. The training should minimize the risk functional [3]

$$R = \frac{1}{2N} \sum \left(d_j - F(x_j)\right)^2 \tag{3.8}$$

where d_j is the desired output pattern for the prototype vector x_j, ((.)) is the Euclidean norm of the enclosed vector, and N is the total number of samples presented to the network in training.

The decision rule therefore can be given by the output of the network

$$y_{kj} = F_k(x_j) \tag{3.9}$$

for the jth input vector x_j.

The MLPs provide global approximations to nonlinear mapping from the input to the output layers. As a result, MLPs are able to generalize in the regions of the input space where little or no data are available [3]. The size of the network is an important consideration from both the performance and computational points of view.

3.4 Multi Layer Perceptron

In a layered ANN, the neurons are organized in the form of layers. In this simplest form of layered network, there is an input layer of source nodes that projects onto an output layer of neurons, but not vice versa. In other words, this network is strictly a feedforward or acyclic type. Multilayer Perceptron distinguishes itself by the presence of one or more hidden layers, whose computation nodes are correspondingly called hidden neurons or hidden units. The function of neurons is to intervene between the external input and the network output in some useful manner. By adding one or more hidden layers, the network is enabled to extract higher-order statistics. The ability of hidden neurons to extract higher-order statistics is particularly valuable when the size of the input layer is large [3]. The source nodes in the input layer of the network supply respective elements of the activation pattern, which constitute the input signals applied to the neurons (computation nodes) in the second layer. The output signals of the second layer are used as inputs to the third layer, and so on for the rest of the network [3]. The set of output signals of the neurons in the output or final layer of the network constitutes the nodes in the input (first) layer. The architectural graph in Fig. 3.5 shows a representation of a simple 3-layer feedforward ANN with 4 inputs, 5 hidden nodes, and 1 output [3].

The MLP is the product of several researchers like Frank Rosenblatt (1958), H. D. Block (1962), M. L. Minsky with S. A. Papart (1988). Backpropagation, the training algorithm, was discovered independently by several researchers [Rumelhart et al. (1986) and also McClelland and Rumelhart (1988)].

David E. Rumelhart, known as the Godfather of Connectionism, was a cognitive scientist who exploited a wide range of formal methods to address issues and topics in cognitive science. Connectionism is an approach to cognitive science that characterizes learning and memory through the discrete interactions between specific

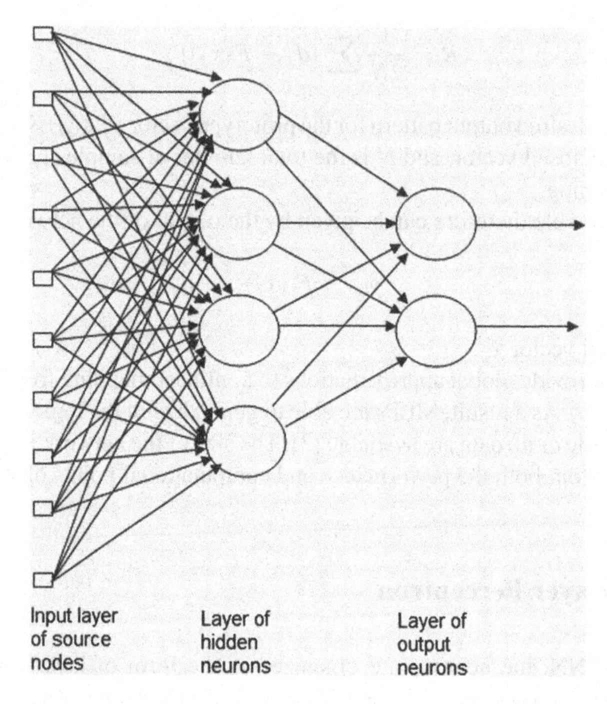

Fig. 3.5 Multilayer feedforward network

nodes of vast ANNs. Rumelhart and McCelland's two-volume book parallel distributed processing (PDP) provided an overall framework in which multilayer ANNs and the backpropagation algorithm made perfect sense. It also promoted the idea of distributed coding where the user seeks a coding that is not efficient in the usual senses but on which makes learning easier.

A simple perceptron is a single McCulloch-Pitts neuron trained by the perceptron learning algorithm is given as

$$O_x = g\left([w] \cdot [x] + b\right) \tag{3.10}$$

where $[x]$ is the input vector, $[w]$ is the associated weight vector, b is a bias value, and $g(x)$ is the activation function. Such a setup, namely the perceptron, is able to classify only linearly separable data. A MLP, in contrast, consists of several layers of neuron (Fig. 3.5). The expression for output in a MLP with one hidden layer is given as

$$O_x = \sum_{i=1}^{N} \beta_i g[w]_i \cdot [x] + b_i \tag{3.11}$$

where β_i is the weight value between the ith hidden neuron.

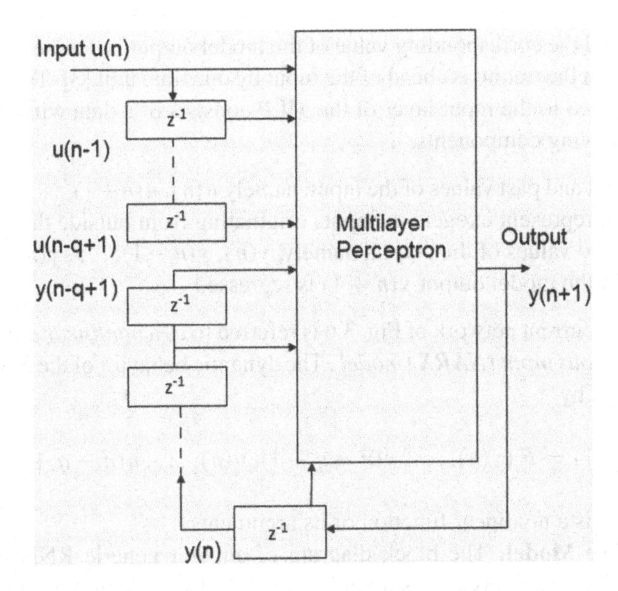

Fig. 3.6 Nonlinear autoregressive ANN with exogenous inputs model

The process of adjusting the weights and biases of a perceptron or MLP is known as *training*. The perceptron algorithm for training simple perceptrons consists of comparing the output of the perceptron with an associated target value. The most common training algorithm for MLPs is (error) backpropagation. This algorithm entails a BP of the error correction through each neuron in the network in the preceding layers.

3.5 Recurrent Neural Network

Recurrent Neural Network (RNN)s are ANNs with one or more feedback loops. The feedback can be of a local or global kind. The RNN maybe considered to be an MLP having a local or global feedback in a variety of forms. It may have feedback from the output neurons of the MLP to the input layer. Yet another possible form of global feedback is from the hidden neurons of the ANN to the input layer [3].

The architectural layout of a recurrent network takes many different forms. Four specific network architectures are commonly used, each of which have a specific form of global feedback.

1. **Input-Output Recurrent Model**: The architecture of a generic recurrent network that follows naturally from a MLP is shown in Fig. 3.6. The model has a single input that is applied to a tapped-delay-line memory of q units. It has a single output that is fed back to the input via another tapped-delay-line memory also using q units. The contents of these two tapped-delay-line memories are used to feed the input layer of the MLP. The present value of the model input is denoted

by $u(n)$, and the corresponding value of the model output is denoted by $y(n+1)$. It means that the output is ahead of the input by one time unit [3]. Thus, the signal vector applied to the input layer of the MLP consists of a data window made up of the following components.

- Present and past values of the input, namely $u(n), u(n+1), \ldots, u(n-q+1)$ which represent exogenous inputs originating from outside the network.
- Delayed values of the output, namely, $y(n), y(n-1), \ldots, y(n-q+1)$, on which the model output $y(n+1)$ is regressed.

Thus, the recurrent network of Fig. 3.6 is referred to as a *nonlinear autoregressive with exogenous input (NARX) model*. The dynamic behavior of the **NARX** model is described by

$$y(n+1) = F\left(y(n), \ldots, y(n-q+1), u(n), \ldots, u(n-q+1)\right) \quad (3.12)$$

where $F(.)$ is a nonlinear function of its arguments.

2. **State-Space Model**: The block diagram of another generic RNN represented using a state-space model is shown in Fig. 3.7. The hidden neurons define the state of the network. The output of the hidden layer is fed back to the input layer via a bank of unit delays. The input layer consists of a concatenation of feedback nodes and source nodes. The network is connected to the external environment via the source nodes. The number of unit delays used to feed the output of the hidden layer back to the input layer determines the order of the model. Let the m-by-1 vector $u(n)$ denote the input vector, and the q-by-1 vector $x(n)$ denote the output of the hidden layer at time n [3]. The dynamic behavior of the model in Fig. 3.7 can be described by the pair of Eqs. (3.13) and (3.14)

$$x(n+1) = f(x(n), u(n)) \quad (3.13)$$

$$y(n) = Cx(n) \quad (3.14)$$

where $f(.,.)$ is a nonlinear function characterizing the hidden layer, and **C** is the matrix of synaptic weights characterizing the output layer. The hidden layer is nonlinear, but the output layer is linear.
The state-space model includes several recurrent architectures as special cases. As for example, the simple recurrent network (SRN) described by Elman (1990) is depicted Fig. 3.8. Elman's network has an architecture similar to that of Fig. 3.7 expect for the fact that the output layer may be nonlinear and the bank of unit delays at the outputs is omitted.

The basic algorithms for RNN training can be summarized as below:

1. **Backpropagation Through Time**: The backpropagation through time (BPTT) algorithm for training of a RNN is an extension of the standard BP algorithm. It may be derived by unfolding the temporal aspect of the structure into a layered

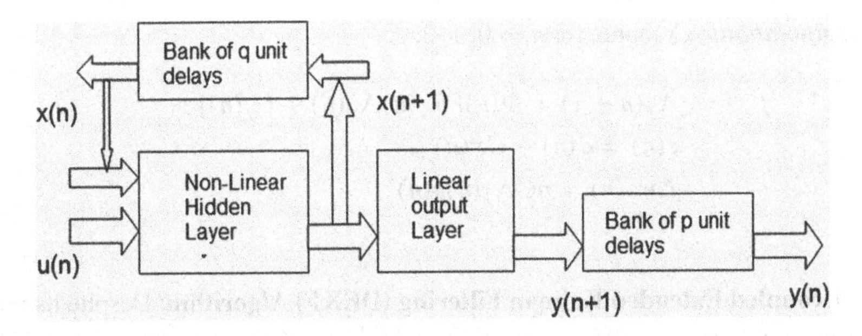

Fig. 3.7 State-space model

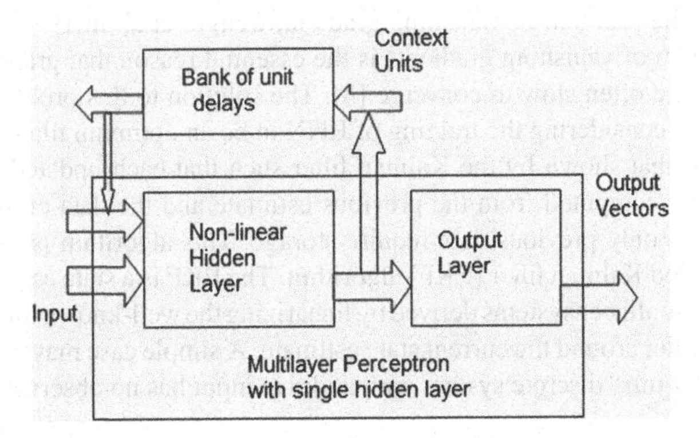

Fig. 3.8 Elman's network

feedforward network, the topology of which grows by one layer at every time step [3].

2. **Real-Time Recurrent Learning**: Another learning method referred to as *Real-Time Recurrent Learning* (RTRL) derives its name from the fact that adjustments are made to the synaptic weights of a fully connected recurrent network in real time. The RTRL algorithm can be summarized as follows [3]:
   Parameters: Let m be dimensionality of input space, q be dimensionality of state space, p be dimensionality of output space, and w_j be synaptic weight vector of neuron.
   Initialization:
   (a) Set the synaptic weights of the algorithm to small values selected from a uniform distribution.
   (b) Set the initial value of the state vector $x(0) = 0$.
   (c) Let $\Lambda_j(n)$ be a $q - by - (q + m + 1)$ matrix defined as the partial derivative of the state vector $x(n)$ with respect to the weight vector w_j. Set $\Lambda_j(0) = 0$ $for\ j = 1, 2, \ldots, q$.

Computations: Compute for $n = 0, 1, 2, \ldots,$

$$\Lambda_j(n+1) = \Phi(n)[W_a(n)\Lambda_j(n) + U_j(n)]$$
$$e(n) = d(n) - Cx(n)$$
$$\Delta w_j(n) = \eta C \Lambda_j(n) e(n)$$

3. **Decoupled Extended Kalman Filtering (DEKF) Algorithm**: Despite its relative simplicity (as compared to BPTT and RTRL), it has been demonstrated that one of the difficulties of using direct gradient descent algorithms for training RNNs is the problem of vanishing gradient. Bengio et al. in 1994 showed that the problem of vanishing gradients is the essential reason that gradient descent methods are often slow to converge [3]. The solution to this problem has been derived by considering the training of RNN to be an optimum filtering problem similar to that shown by the Kalman filter such that each updated estimate of the state is computed from the previous estimate and the data currently available. Here, only previous data require storage. This algorithm is derived from the extended Kalman filter (EKF) algorithm. The EKF is a state estimation technique for nonlinear systems derived by linearizing the well-known linear-systems Kalman filter around the current state estimate. A simple case maybe considered such that a time-discrete system with additive input has no observation noise:

$$x(n+1) = f(x(n)) + q(n) \tag{3.15}$$
$$d(n) = h_n(x(n)) \tag{3.16}$$

where $x(n)$ is the system's internal state vector, $f(.)$ is the system's state update function (linear in original Kalman filters), $q(n)$ is external input to the system (an uncorrelated Gaussian white noise process, can also be considered as process noise), $d(n)$ is the system's output, and h_n is a time-dependent observation function (also linear in the original Kalman filter). At time $n = 0$, the system state $x(0)$ is guessed by a multidimensional normal distribution with mean $\hat{x}(0)$ and covariance matrix $P(0)$. The system is observed until time n through $d(0),\ldots,$ $d(n)$. The task addressed by the extended Kalman filter is to give an estimate $\hat{x}(n+1)$ of the true state $x(n+1)$, given the initial state guess and all previous output observations. This task is solved by the following two time update (Eqs. 3.17, 3.18) and three measurement update computations (Eqs. 3.19–3.21):

$$\hat{x}^*(n) = F\left(\hat{x}(n)\right) \tag{3.17}$$
$$P^*(n) = F(n)P(n-1)F(n)^t + Q(n) \tag{3.18}$$

and

$$K(n) = P^*(n)H(n)\left[H(n)^t P^*(n)H(n)\right]^{-1} \tag{3.19}$$

$$\hat{x}^*(n+1) = \hat{x}^*(n) + K(n)\xi(n) \tag{3.20}$$

$$P(n+1) = P^*(n) - K(n)H(n)^t P^*(n) \tag{3.21}$$

Here, $F(n)$ and $H(n)$ are the Jacobians as per the notation used by Singhal and Wu in 1989 of the components of f, h_n with respect to the state variables, evaluated at the previous state estimate

$$\xi(n) = d(n) - h_n\left(\hat{x}(n)\right) \tag{3.22}$$

where, $\xi(n)$ is the error (difference between observed output and output calculated from state estimate $\hat{x}(n)$, $P(n)$ is an estimate of the conditional error covariance matrix, $Q(n)$ is the (diagonal) covariance matrix of the process noise, and the time updates $\hat{x}^*(n)$, $P^*(n)$ of state estimate and state error covariance estimate are obtained from extrapolating the previous estimates with the known dynamics f. Information contained in $d(n)$enters the measurement update in the form of $\xi(n)$ and is accumulated in the Kalman gain $K(n)$. In the case of classical (linear, stationary) Kalman filtering, $F(n)$ and $H(n)$ constant, and the state estimates converge to the true conditional mean state. For nonlinear $f(.)$ and h_n, usually does not hold. The application of extended Kalman filters generates locally optimal state estimates [5].

- **Application of EKF for RNN Training**: Assuming that a RNN perfectly reproduces the input output training data with very little imperfection and shows a time series dependence like

$$u(n) = [u_1(n), \ldots, u_N(n)]^{\mathrm{T}},\ d(n) = [d_1(n), \ldots, d_N(n)]^{\mathrm{T}}, \quad n = 1, 2, \ldots \mathrm{T} \tag{3.23}$$

The output $d(n)$ of the RNN is passed through a function $h(.)$ with products of the weights $\mathbf{w}$ and input $\mathbf{u}$ up to n is expressed as

$$d(n) = h[w, u(0), u(1), \ldots, u(n)] \tag{3.24}$$

Assuming the presence of uncorrelated Gaussian noise $q(n)$, the network weight update expression maybe expressed as

$$w(n+1) = w(n) + q(n) \tag{3.25}$$

$$d(n) = h_n(w(n)) \tag{3.26}$$

The measurement updates become

$$K(n) = P(n)H(n)\left[H(n)^t P(n)H(n)\right]^{-1} \tag{3.27}$$

$$\hat{w}(n+1) = \hat{w}(n) + K(n)\xi(n) \tag{3.28}$$

$$\text{and } P(n+1) = P(n) - K(n)H(n)^t P(n) + Q(n) \tag{3.29}$$

A learning rate η to compensate for initially bad estimates of $P(n)$ can be introduced into the Kalman gain update expression given as

$$K(n) = P(n)H(n)\left[\left(\frac{1}{\eta}\right)I + H(n)^t P(n)H(n)\right]^{-1} \tag{3.30}$$

Inserting process noise $q(n)$ into EKF has been claimed to improve the algorithm's numerical stability, and to avoid getting stuck in poor local minima in Puskorius and Feldkamp 1994 [3, 5, 6]. EKF is a second-order gradient descent algorithm. As a result it uses curvature information of the (squared) error surface. Due to exploiting second order curvature for linear noise-free systems the Kalman filter can converge in a single step which makes RNN training faster [3, 5, 6].
EKF requires the derivatives *H(n)* of the network outputs with respect to the weights evaluated at the current weight estimate. These derivatives can be exactly computed as in the RTRL algorithm, at cost $O(N^4)$.

- **Decoupled Extended Kalman Filter (DEKF) and RNN Learning**: Over and above of the calculation of $H(n)$, the update of $P(n)$ is computationally most expensive in EKF and it requires $O(LN^2)$ operations. By configuring the network with decoupled subnetworks, a block-diagonal $P(n)$ can be obtained which results in considerable reduction in computations as proved by Feldkamp et al. in 1998 [5]. This gives rise to the DEKF algorithm which can be summarized as below [3]:
  Initialization:
  (a) Set the synaptic weights of the RNN to small values selected from a uniform distribution.
  (b) From the diagonal elements of the covariance matrix $\mathbf{Q}(n)$ (process noise between 10^{-6} to 10^{-2}).
  (c) Set $\mathbf{K}(1, 0) = \delta^{-1}\mathbf{I}$, where δ is a small positive constant.
  Computation: For $n = 1, 2, \ldots$ compute

$$\Gamma(n) = \left[\sum_{i=1}^{g} \mathbf{C}_i(n)\mathbf{K}_i(n, n-1)\mathbf{C}_i^{\mathrm{T}}(n) + \mathbf{R}(n)\right]^{-1} \tag{3.31}$$

$$\mathbf{G}_i(n) = \mathbf{K}_i(n, n-1)\mathbf{C}_i^{\mathrm{T}}(n)\Gamma(n) \tag{3.32}$$

4. **Complex-Valued RTRL Algorithm**: Figure 3.9 shows an CFRNN, which consists of N neurons with p external inputs. The network has two distinct layers consisting of the external input-feedback layer and a layer of processing elements. Let $y_l(k)$ denote the complex-valued output of each neuron, $l = 1, \ldots, N$

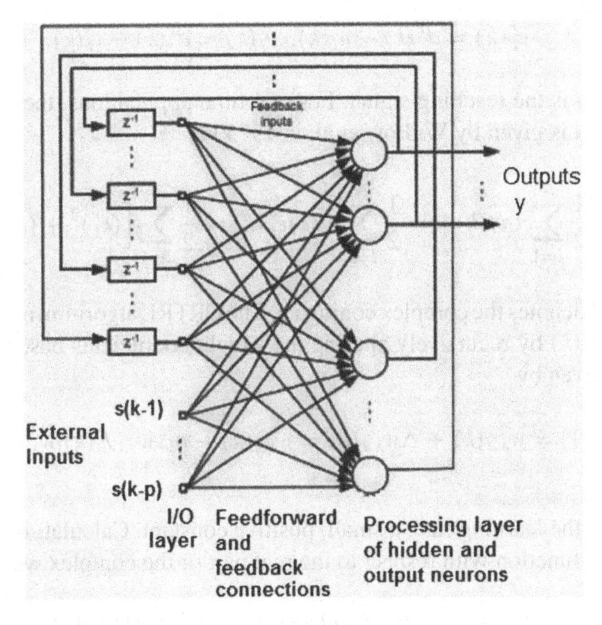

Fig. 3.9 Fully connected RNN

at time index k and $s(k)$ the $(1 \times p)$ external complex-valued input vector. The overall input to the network $I(k)$ represents the concatenation of vectors $y(k)$, $s(k)$, and the bias input $(1 + j)$, and is given by

$$I(k) = [s(k-1), \ldots, s(k-p), 1+j, y_1(k-1), \ldots, y_N(k-1)]^{\mathrm{T}} \quad (3.33)$$
$$= I_n^r(k) + jI_n^i(k), \quad n = 1, \ldots, p+N+1 \quad (3.34)$$

where $j = \sqrt{-1}$, $(.)^{\mathrm{T}}$ denotes the vector transpose operator, and the superscripts $(.)^r$ and $(.)^p$ denote, respectively, the real and imaginary parts of a complex number. A complex-valued weight matrix of the network is denoted by W, where for the ith neuron, its weights form a $(p+F+1) \times 1$ dimensional weight vector $W_l = [w_{l,1,\ldots}, w_{l,p+F+1}]^{\mathrm{T}}$ where F is the number of feedback connections. The feedback connections represent the delayed output signals of the CFRNN.
For simplicity, it can be stated that

$$y_l(k) = \phi^r(\mathrm{net}_l(k)) + j\phi^i(\mathrm{net}_l(k)) = u1(k) + jv_l(k) \quad (3.35)$$

$$\mathrm{net}_l(k) = \sigma_l(k) + j\tau_l(k) \quad (3.36)$$

where $\phi(.)$ is a complex nonlinear activation function.
The output error consists of its real and imaginary parts and is defined as

$$e_l(k) = d(k) - y_l(k) = e_l^r(k) + je_l^i(k) \quad (3.37)$$

$$e_l^r(k) = d^r(k) - u_l(k),\ e_l^i(k) = d^i(k) - v_l(k), \tag{3.38}$$

where $d(k)$ is the teaching signal. For real-time applications, the cost function of the RNN is given by Widrow et al. in 1975 [7],

$$E(k) = \frac{1}{2}\sum_{l=1}^{N} | e_l(k) |^2 = \frac{1}{2}\sum_{l=1}^{N} e_l(k)e_l^*(k) = \frac{1}{2}\sum_{l=1}^{N}\left[(e_l^r)^2 + (e_l^i)^2\right], \tag{3.39}$$

where (.)* denotes the complex conjugate. The CRTRL algorithm minimizes cost function $E(k)$ by recursively altering the weight coefficients based on gradient descent, given by

$$w_{s,t}(k+1) = w_{s,t}(k) + \Delta w_{s,t}(k) = w_{s,t}(k) - \eta\Delta w_{s,t}E(k)_{|}w_{s,t} = w_{s,t}(k), \tag{3.40}$$

where η is the learning rate, a small, positive constant. Calculating the gradient of the cost function with respect to the real part of the complex weight gives

$$\frac{\partial E(k)}{\partial w_{s,t}^r(k)} = \frac{\partial E}{\partial u_l}\left(\frac{\partial u_l(k)}{\partial w_{s,t}^r(k)}\right) + \frac{\partial E}{\partial v_l}\left(\frac{\partial v_l(k)}{\partial w_{s,t}^r(k)}\right),$$
$$1 \le l, s \le N, 1 \le t \le p+N+1 \tag{3.41}$$

Similarly, the partial derivative of the function with respect to the imaginary part of the complex weight yields

$$\frac{\partial E(k)}{\partial w_{s,t}^i(k)} = \frac{\partial E}{\partial u_l}\left(\frac{\partial u_l(k)}{\partial w_{s,t}^i(k)}\right) + \frac{\partial E}{\partial v_l}\left(\frac{\partial v_l(k)}{\partial w_{s,t}^i(k)}\right),$$
$$1 \le l, s \le N, 1 \le t \le p+N+1 \tag{3.42}$$

The factors $\frac{\partial y_l(k)}{\partial w_{s,t}^r k} = \frac{\partial u_l(k)}{\partial w_{s,t}^r(k)} + j\frac{\partial v_l(k)}{\partial w_{s,t}^r(k)}$ and $\frac{\partial y_l(k)}{\partial w_{s,t}^i k} = \frac{\partial u_l(k)}{\partial w_{s,t}^i(k)} + j\frac{\partial v_l(k)}{\partial w_{s,t}^i(k)}$ are measures of sensitivity of the output of the mth unit at time k to a small variation in the value of $w_{s,t}(k)$. This sensitivities can be evaluated as

$$\frac{\partial u_l(k)}{\partial w_{s,t}^r(k)} = \frac{\partial u_l}{\partial \sigma_l} \cdot \frac{\partial \sigma_l}{\partial w_{s,t}^r(k)} + \frac{\partial u_l}{\partial \tau_l} \cdot \frac{\partial \tau_l}{\partial w_{s,t}^r(k)} \tag{3.43}$$

$$\frac{\partial u_l(k)}{\partial w_{s,t}^i(k)} = \frac{\partial u_l}{\partial \sigma_l} \cdot \frac{\partial \sigma_l}{\partial w_{s,t}^i(k)} + \frac{\partial u_l}{\partial \tau_l} \cdot \frac{\partial \tau_l}{\partial w_{s,t}^i(k)} \tag{3.44}$$

$$\frac{\partial v_l(k)}{\partial w_{s,t}^r(k)} = \frac{\partial v_l}{\partial \sigma_l} \cdot \frac{\partial \sigma_l}{\partial w_{s,t}^r(k)} + \frac{\partial v_l}{\partial \tau_l} \cdot \frac{\partial \tau_l}{\partial w_{s,t}^r(k)} \tag{3.45}$$

Table 3.1 Computational complexity of RNN training algorithms

Sl Num	Algorithm	Computational complexity
1	BPTT	Time and storage, O(WL + SL)
2	RTRL	Time = $O(WS^2L)$, Storage = $O(WS)$
3	DEKF	Time$Time = O(p^2W + p\sum_{i=1}^{g} k_i^2)$ Storage = $O(\sum_{i=1}^{g} k_i^2)$ p is the number of outputs g is the number of groups k_i is the number of neurons in group k_i
4	CRTL	Time complexity = $O(N^4)$ Storage complexity = $O(N^3)$

$$\frac{\partial v_l(k)}{\partial w_{s,t}^i(k)} = \frac{\partial v_l}{\partial \sigma_l} \cdot \frac{\partial \sigma_l}{\partial w_{s,t}^i(k)} + \frac{\partial v_l}{\partial \tau_l} \cdot \frac{\partial \tau_l}{\partial w_{s,t}^i(k)} \tag{3.46}$$

5. **Computational Complexity of BPTT, RTRL, DEKF, and CRTRL Algorithms**: Let S be the number of states, W be the number of synaptic weights, and L be the length of the training sequence. The computational complexity [3] of these algorithms maybe summarized as in Table 3.1.

3.6 Probabilistic Neural Network

The Probabilistic Neural Network (PNN), introduced by Donald Specht in 1988, is a 3-layer, feedforward, one pass training algorithm used for classification and mapping of data. It is based on well-established statistical principles derived from Bayes decision strategy and nonparametric kernel-based estimators of probability density functions (pdf). The basic operation performed by the PNN is an estimation of the probability density function of features of each class from the provided training samples using Gaussian kernel. These estimated densities are then used in a Bayes decision rule to perform the classification. An advantage of the PNN is that it is guaranteed to approach the Bayes optimal decision surface provided that the class probability density functions are smooth and continuous.

The PNN is closely related to Parzen window probability density function estimator. A PNN consists of several subnetworks, each of which is a Parzen window pdf estimator for each of the classes. The input nodes are the set of measurements. The second layer consists of the Gaussian functions formed using the given set of data points as centers. The third layer performs an average operation of the outputs from the second layer for each class. The fourth layer performs a vote, selecting the largest value. The associated class label is then determined.

PNN is found to be an excellent pattern classifier, outperforming other classifiers including BP ANNs. However, it is not robust with respect to affine transformations of feature space, and this can lead to poor performance on certain data [8].

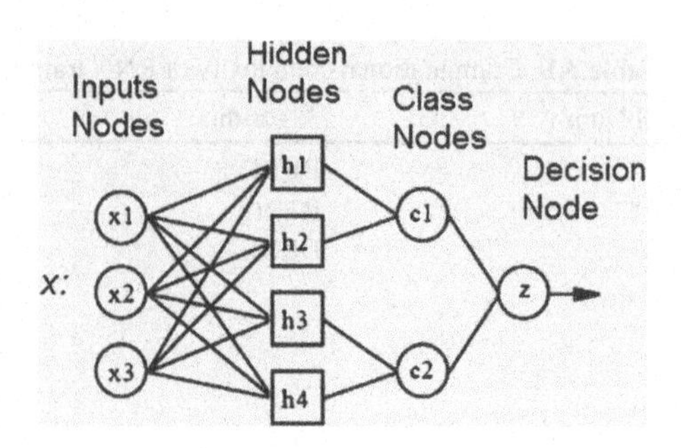

Fig. 3.10 Architecture of probabilistic neural network (PNN) [10]

A PNN is defined as an implementation of statistical algorithm called kernel discriminate analysis in which the operations are organized into multilayered feed-forward network with four layers: input layer, pattern layer, summation layer, and output layer. A PNN is predominantly a classifier since it can map any input pattern to a number of classifications. Among the main advantages that discriminate PNN are

1. Fast training process,
2. An inherently parallel structure,
3. Guaranteed to converge to an optimal classifier as the size of the representative training set increases, and
4. Training samples can be added or removed without extensive retraining.

Accordingly, a PNN learns more quickly than many other ANN models and have had success on a variety of applications. Based on these facts and advantages, PNN can be viewed as a supervised ANN that is capable of using it in system classification and pattern recognition [9].

3.6.1 Architecture of a PNN Network

All PNN networks have four layers as shown in Fig. 3.10:

1. **Input layer**: There is one neuron in the input layer for each predictor variable. In the case of categorical variables, $N - 1$ neurons are used where N is the number of categories. The input neurons or processing before the input layer standardizes the range of the values by subtracting the median and dividing by the interquartile range. The input neurons then feed the values to each of the neurons in the hidden layer [10].
2. **Hidden layer**: This layer has one neuron for each case in the training data set. The neuron stores the values of the predictor variables for the case along with the target value. When presented with the x vector of input values from the input layer, a hidden neuron computes the Euclidean distance of the test case from the neurons center point and then applies the RBF kernel function using the sigma values. The resulting value is passed to the neurons in the pattern layer [10].

3. **Pattern layer or Summation layer**: The next layer in the network is one pattern neuron for each category of the target variable. The actual target category of each training case is stored with each hidden neuron; the weighted value coming out of a hidden neuron is fed only to the pattern neuron that corresponds to the hidden neurons category. The pattern neurons add the values for the class they represent. Hence, it is a weighted vote for that category [10].
4. **Decision layer**: For PNN, the decision layer compares the weighted votes for each target category accumulated in the pattern layer and uses the largest vote to predict the target category [10].

The development of the PNN relies on Parzen windows classifiers. The Parzen windows method is a nonparametric procedure that synthesizes an estimate of a pdf by superposition of a number of windows, replicas of a function (often the Gaussian). The Parzen windows classifier takes a classification decision after calculating the pdf of each class using the given training examples [11]. The multicategory classifier decision is expressed as follows

$$p_k f_k > p_j f_j, \quad \text{for all } j \neq k \tag{3.47}$$

where p_k is the prior probability of occurrence of examples from class k, and f_k is the estimated pdf of class k. The calculation of the pdf is performed with the following algorithmic steps.

Add up the values of the d-dimensional Gaussians, evaluated at each training example, and scale the sum to produce the estimated probability density [11].

$$f_k(x) = \frac{1}{2 \times \pi} \frac{d}{2} \sigma^d \left(\frac{1}{N}\right) \sum_{i=l}^{NK} \exp\left[\frac{-(x - x_{ki})^{\mathrm{T}}(x - x_{ki})}{(2 \times \sigma^2)}\right] \tag{3.48}$$

where x_{ki} is the d-dimensional ith example from class k. This function $f_k(x)$ is a sum of small multivariate Gaussian probability distributions centered at each training example. Using probability distributions allows to achieve generalization beyond the provided examples. The index k in $f_k(x)$ indicates difference in spread between the distributions of the classes. As the number of the training examples and their Gaussians increases, the estimated pdf approaches the true pdf of the training set [11]. The classification decision is taken according to the inequality:

$$\sum_{i=l}^{Nk} \exp\left[\frac{-(x - x_{ki})^T(x - x_{ki})}{(2 \times \sigma^2)}\right] > \sum_{i=l}^{Nj} \exp\left[\frac{-(x - x_{ji})^T(x - x_{ji})}{(2 \times \sigma^2)}\right] \text{ for all } j \neq k \tag{3.49}$$

which is derived using prior probabilities calculated as the relative frequency of the examples in each class

$$P_k = N_k \div N \tag{3.50}$$

where N is the number of all training examples, and N_k is the number of examples in class k. The Parzen windows classifier uses the entire training set of examples to perform classification. It means that it requires storage of the training set in the computer memory. The speed of computation is proportional to the training set size [11].
As mentioned earlier, the PNN is a direct continuation of the work on Bayes classifiers. The PNN learns to approximate the pdf of the training examples. More precisely, the PNN is interpreted as a function which approximates the probability density of the underlying examples distribution [11]. The PNN consists of nodes allocated in three layers after the inputs. There is one pattern node for each training example. Each pattern node forms a product of the weight vector and the given example for classification, where the weights entering a node are from a particular example. After that, the product is passed through the activation function given as

$$\exp\left[\frac{(x^{\mathrm{T}}W_{ki}-1)}{\sigma^2}\right] \tag{3.51}$$

where x^{T} is the pattern vector, W_{ki} is the ith weigth value in the kth layer of weight matrix [w] and σ is the smoothing parameter. Each summation node receives the outputs from pattern nodes associated with a given class:

$$\sum_{i=l}^{Nk}\exp\left[\frac{(x^{\mathrm{T}}w_{ki}-1)}{\sigma^2}\right] \tag{3.52}$$

The output nodes are binary neurons that produce the classification decision

$$\sum_{i=l}^{Nk}\exp\left[\frac{(x^{\mathrm{T}}w_{ki}-1)}{\sigma^2}\right] > \sum_{i=l}^{Nj}\exp\left[\frac{(x^{\mathrm{T}}w_{kj}-1)}{\sigma^2}\right] \tag{3.53}$$

The only factor that needs to be selected for training is the smoothing factor, which represents the deviation of the Gaussian functions. It is governed by the following considerations:

1. Too small deviations cause a very spiky approximation which cannot generalize well and
2. Too large deviations smooth out details.

An appropriate deviation is chosen by experiment [11].

3.7 Self-Organizing Map

Self-Organizing Feature Maps (SOFM)s or Self Organizing Map (SOM) are a type of ANN develop by Kohonen [12]. SOM (Fig. 3.11) reduces dimensions by producing a map of usually one or two dimensions that plot the similarities of the data by

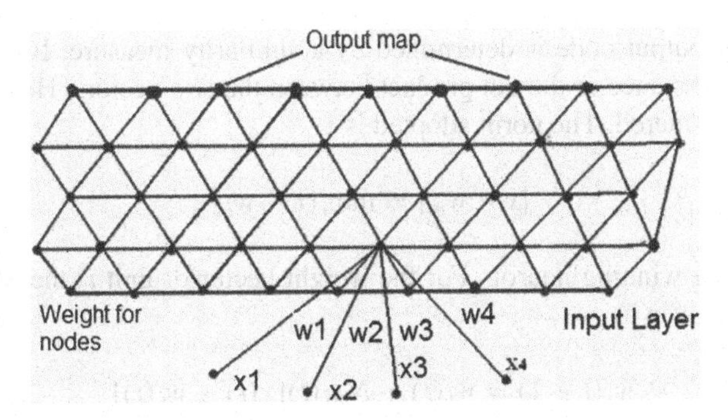

Fig. 3.11 SOFM

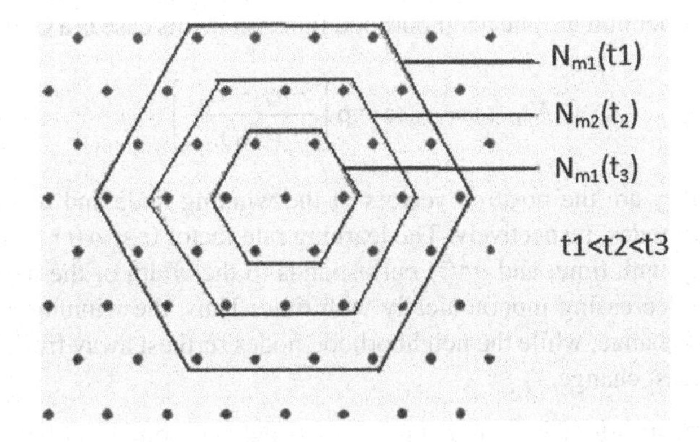

Fig. 3.12 Topological neighborhood of SOFM

grouping the similar data items together. SOMs accomplish two things. They reduce dimensions and display similarities. An input variable is presented to the entire network of neurons during training. The neuron whose weight most closely matches that of the input variable is called the winning neuron, and this neuron's weight is modified to more closely match the incoming vector. The SOM uses an unsupervised learning method to map high-dimensional data into a 1-D, 2-D, or 3-D data space, subject to a topological ordering constraint. A major advantage is that the clustering produced by the SOM retains the underlying structure of the input space, while the dimensionality of the space is reduced. As a result, a neuron map is obtained with weights encoding the stationary probability density function $p(X)$ of the input pattern vectors [13].

- **Network Topology**: The SOM consists of two main parts, the input layer and the output map. There is no hidden layer. The dimensionality of the input layer is not restricted, while the output map has dimensionality 1-D, 2-D, or 3-D. An example of a SOFM is shown in Fig. 3.11, and a planar array of neurons with the neighborhoods is shown in Fig. 3.12

The winning output node is determined by a similarity measure. It can either be the Euclidean distance or the dot product between the two vectors. Here, Euclidean distance is considered. The norm adopted is

$$|x - w_m| = \min_i |x - w_i| \tag{3.54}$$

where w_i is the winning neuron. For the weight vector of unit i, the SOM weight update rule is given as

$$w_i(t+1) = w_i(t) + N_{mi}(t)[x(t) - w_i(t)] \tag{3.55}$$

where t is the time, $x(t)$ is an input vector, and $N_{mi}(t)$ is the neighborhood kernel around the winner unit m. The neighborhood function in this case is a Gaussian given by

$$N_{mi}(t) = \alpha(t) \exp\left[\frac{-|r_i - r_m|}{\sigma^2(t)}\right] \tag{3.56}$$

where r_m and r_i are the position vectors of the winning node and of the winning neighborhood nodes, respectively. The learning rate factor $0 < \alpha(t) < 1$ decreases monotonically with time, and $\sigma^2(t)$ corresponds to the width of the neighborhood function also decreasing monotonically with time. Thus, the winning node undergoes the most change, while the neighborhood nodes furthest away from the winner undergo the least change.

- **Batch training**: The incremental process defined above can be replaced by a batch computation version which is significantly faster. The batch training algorithm is also iterative, but instead of presenting a single data vector to the map at a time, the whole data set is given to the map before any adjustments are made. In each training step, the data set is partitioned such that each data vector belongs to the neighborhood set of the map unit to which it is closest, the Voronoi set [13]. The sum of the vectors in each Voronoi set are calculated as:

$$S_i(t) = \sum_{j=1}^{^nV_i} X_j \tag{3.57}$$

where ^nV_i is the number of samples in the Voronoi set unit i. The new values of the weight vector are then calculated as

$$w_i(t+1) = \frac{\sum_{j=1}^{m} N_{ij}(t) S_j(t)}{\sum_{j=1}^{m} (^nV_i) h_{ij}(t)} \tag{3.58}$$

where m is the number of map units, and ^nV_j is the number of sample falling into Voronoi set V_i.

- **Quality Measures**: Although the issue of SOM quality is not a simple one, two evaluation criteria, resolution and topology preservation, are commonly used [14]. Other methods are available in [14, 15].

3.7.1 Competitive Learning and Self-Organizing Map (SOM)

In competitive learning, the neurons in a layer work with the principle 'winners take all.' There is a competition between the neurons to find the closest match to that of the input sample vector. While the competition is on, the connectionist weights update during the process and the neuron with the closest similarity is assigned the input sample. SOMs use competitive learning to perform a host of tasks which includes pattern matching, recognition, and feature mapping [16]. The following are among some of methods used for training and weight updating of SOMs as part of competitive learning [4]:

1. **Inner Product** and
2. **Euclidean Distance-Based Competition**.

A detailed account of the two methods maybe given as below [3, 4]:

1. **Inner Product and Euclidean Distance-Based Competition**: The training process of the SOM is linked with the creation of a codebook which provides the account of finding the best match among the set of input patterns given to it. There are two ways in which the best matching codebook vector can be found. The first way is to employ an inner product criterion—select the best reference vector by choosing the neuron in the competitive layer that receives the maximum activation. This means that for the current input vector X_k, we compute all neuron activations
$$y_j = X_k^{\mathrm{T}} W_j \quad j = 1, \ldots, m \tag{3.59}$$
and the winning neuron index J satisfies
$$y_J = \max_j X_k^{\mathrm{T}} W_j \tag{3.60}$$
Alternatively, one might select the winner based on a Euclidean distance measure. Here, the distance is measured between the present input X_k and the weight vectors W_j, and the winning neuron index J satisfies
$$|X_k - W_J| = \min_j |X_k - W_j| \tag{3.61}$$
It is misleading to think that these methods of selecting the winning neuron are entirely distinct. To see this, assume that the weight *equinorm* property holds for all weight vectors:
$$|W_1| = |W_2| = \cdots = |W_m| \tag{3.62}$$

We now rework the condition given in Eq. (3.61) as follows:

$$|X_k - W_J|^2 = \min_j |X_k - W_j|^2 \tag{3.63}$$

$$\Rightarrow (X_k - W_J)^{\mathrm{T}}(X_k - W_J) = \min_j\{(X_k - W_j)^{\mathrm{T}}(X_k - W_j)\} \tag{3.64}$$

$$\Rightarrow |X_k|^2 - 2X_k^{\mathrm{T}}W_J + |W_J|^2 = \min_j |X_k|^2 - 2X_k^{\mathrm{T}}W_j + |W_j|^2 \tag{3.65}$$

$$\Rightarrow -2X_k^{\mathrm{T}}W_J + |W_J|^2 = \min_j(-2X_k^{\mathrm{T}}W_j + |W_j|^2) \tag{3.66}$$

We now subtract $|W_J|^2$ from both side of Eq. (3.66) and invoke the weight equinorm assumption. This yields:

$$-2X_k^{\mathrm{T}}W_J = \min_j -2X_k^{\mathrm{T}}W_j \tag{3.67}$$

or,

$$X_k^{\mathrm{T}}W_J = \max_j 2X_k^{\mathrm{T}}W_j \tag{3.68}$$

which is the identical to criterion given in Eq. (3.61). Neuron J wins if its weight vector correlates maximally with the impinging input.

2. **A Generalized Competitive Learning Law**: Competitive learning requires that the weight vector of the winning neuron be made to correlate more with the input vector. This is done by perturbation of only the winning weight vector $W_J = (w_{1J}, \ldots, w_{nJ})^{\mathrm{T}}$ toward the input vector. The scalar implementation of this learning law in difference form is presented below as

$$w_{iJ}^{k+1} = w_{iJ}^k + \eta x_i^k \quad i = 1, \ldots, n \tag{3.69}$$

However, as we have seen in such forms of learning, the weights grow without bound and some form of normalization is required. By normalization, we mean that the sum of the instar weight should add up to unity: $\sum_{i=1}^{n} = 1$. This can be ensured by normalizing the inputs within the equation and incorporating an extra weight subtraction term. The complete expression is given as

$$w_{iJ}^{k+1} = w_{iJ}^k + \eta\left(\frac{x_i^k}{\sum_j x_j^k} - w_{iJ}^k\right) \quad i = 1, \ldots, n \tag{3.70}$$

This Eq. (3.70) leads to total weight normalization. To see this, we simply add all n equations (Eq. 3.70) to get the total weight update equation as

$$\tilde{W}_{J(k+1)} = \tilde{W}_{J(k)} + \eta\left(1 - \tilde{W}_{J(k)}\right) \tag{3.71}$$

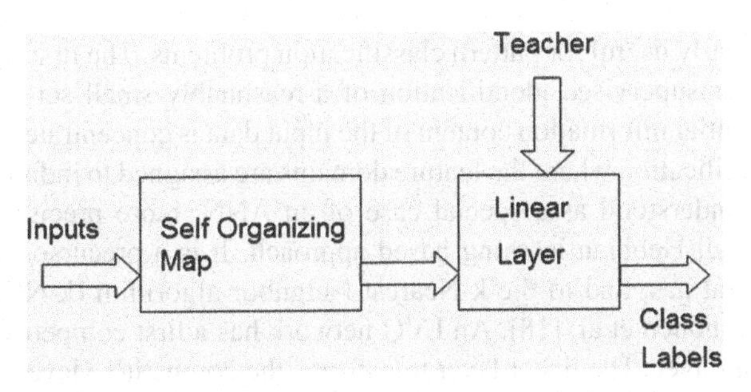

Fig. 3.13 The two stages of LVQ [17]

where $\tilde{W}_{J(k)} = \sum_{i=1}^{n} w_{iJ}^{k}$. It is clear from Eq. (3.71) that the weight sum eventually approaches a point on the unit hyperspher, S^n . If the inputs are already normalized, then normalization is not required within the law and weight update is direct which is given as

$$w_{iJ}^{k+1} = w_{iJ}^{k} + \eta \left(x_i^k - w_{iJ}^{k} \right) \quad i = 1, \ldots, n \tag{3.72}$$

It effectively moves the Jth instar W_J, in the direction of the input vector X_k. We have written this equation for a winning neuron with index J. In general, one may rewrite this equation for all neurons in the field may be expressed as

$$w_{iJ}^{k+1} = \begin{cases} w_{ij}^{k} + \eta(x_i^k - w_{iJ}^{k}), & i = 1, \ldots, n \; j = J \\ w_{ij}^{k} & i = 1, \ldots, n \; j \neq J \end{cases} \tag{3.73}$$

where $J = \arg\max_j \{y_j^k\}$. Based on this, a further generalization can be obtained and write out and a standard competitive form in discrete time be given as

$$w_{ij}^{k+1} = w_{ij}^{k} + \eta s_j^k \left(x_i^k - w_{ij}^{k} \right) \quad i = 1, \ldots, n \; j = 1, \ldots, m \tag{3.74}$$

where $s_j^k = 1$ only for $j = J$ and is zero otherwise, for a hard competitive field. These concepts are relevant for the study of an important class of ANN models that have the foundation in competitive learning. This is described in the following section (Sect. 3.8).

3.8 Learning Vector Quantization

Learning Vector Quantization (LVQ) is a supervised version of vector quantization (VQ). This learning technique uses the class information to reposition the weights or Voronoi vectors slightly, so as to improve the quality of the classifier decision regions. It is a two-stage process, a SOM followed by a linear supervised classifier (Fig. 3.13).

This is particularly useful for pattern classification problems. The first step is feature selection, the unsupervised identification of a reasonably small set of features in which the essential information content of the input data is concentrated. The second step is the classification where the feature domains are assigned to individual classes. LVQ can be understood as a special case of an ANN; more precisely, it applies a winner-take-all Hebbian learning based approach. It is a precursor to SOM and related to neural gas, and to the k-Nearest Neighbor algorithm (k-NN). LVQ was invented by Kohonen et al. [18]. An LVQ network has a first competitive layer and a second linear layer. The linear layer transforms the competitive layers classes into target classifications defined by the user. The classes learned by the competitive layer are referred to as subclasses and the classes of the linear layer as target classes. Both the competitive and linear layers have one neuron per (sub or target) class. Thus, the competitive layer can learn up to S1 subclasses. These, in turn, are combined by the linear layer to form S2 target classes (S1 is always larger than S2) [1].

3.8.1 LVQ Approach

The basic LVQ approach is quite intuitive. It is based on a standard trained SOM with input vectors x and weights/Voronoi vectors w_j [17].

The new factor is that the input data points have associated class information. This allows us to use the known classification labels of the inputs to find the best classification label for each w_j, i.e., for each Voronoi cell . For example, by simply counting up the total number of instances of each class for the inputs within each cell [17].

Then, each new input without a class label can be assigned to the class of the Voronoi cell it falls within [17].

The problem with this is that, in general, it is unlikely that the Voronoi cell boundaries will match up with the best possible classification boundaries, so the classification generalization performance will not be as good as possible. The obvious solution is to shift the Voronoi cell boundaries, so they better match the classification boundaries [17].

3.8.2 LVQ Algorithm

The LVQ is an algorithm for learning classifiers from labeled data samples. Instead of modeling the class densities, it models the discrimination function defined by the set of labeled codebook vectors and the nearest neighborhood search between the codebook and data. In classification, a data point x_i is assigned to a class according to the class label of the closest codebook vector. The training algorithm involves an iterative gradient update of the winner unit.

The following three LVQ algorithms are presented here.

1. **LVQ1**: Assume that a number of 'codebook vectors' m_i (free parameter vectors) are placed into the input space to approximate various domains of the input vector x by their quantized values. Usually, several codebook vectors are assigned to each class of x values, and x is then decided to belong to the same class to which the nearest m_i belongs [18]. Let

$$C = \arg\min_i |x - m_i| \tag{3.75}$$

define the nearest m_i to x, denoted by m_C.
Values for the m_i that approximately minimize the misclassification errors in the above nearest-neighbor classification can be found as asymptotic values in the following learning process. Let $x(t)$ be a sample of input and let the $m_i(t)$ represent sequences of the m_i in the discrete-time domain. Starting with properly defined initial values, the following equations define the basic LVQ1 process [18].

$$\begin{aligned} &m_c(t+1) = m_c(t) + \alpha(t)[x(t) - m_c(t)] \\ &\text{if } x \text{ and } m_c \text{ belong to the same class} \\ &m_c(t+1) = m_c(t) - \alpha(t)[x(t) - m_c(t)] \\ &\text{if } x \text{ and } m_c \text{ belong to the different class} \\ &m_i(t+1) = m_i(t) \end{aligned} \tag{3.76}$$

Here, $0 < \alpha(t) < 1$ and $\alpha(t)$ may be constant or decrease monotonically with time. In the above basic LVQ1, it is recommended that α should initially be smaller than 0.1.

2. **LVQ2.1**: The classification decision in this algorithm is identical with that of the LVQ1. In learning, however, two codebook vectors, m_i and m_j that are the nearest neighbors to x, are now updated simultaneously. One of them must belong to the correct class and the other to a wrong class, respectively. Moreover, x must fall into a zone of values called 'window,' which is defined around the midplane of m_i and m_j. Assume that d_i and d_j are the Euclidean distances of x from m_i and m_j, respectively. Then, x is defined to fall in a 'window' of relative width w if

$$\min\left(\frac{d_i}{d_j}, \frac{d_j}{d_i}\right) > S \quad \text{where } s = \frac{1-w}{1+w} \tag{3.77}$$

A relative 'window' width w of 0.2 to 0.3 is recommendable. The related algorithm is expressed as

$$\begin{aligned} m_i(t+1) &= m_i(t) - \alpha(i)[x(t) - m_i(t)] \\ m_j(t+1) &= m_j(t) + \alpha(t)[x(t) - m_j(t)] \end{aligned} \tag{3.78}$$

where m_i and m_j are the two closest codebook vectors to x, whereby x and m_j belong to the same class, while x and m_i belong to different classes, respectively. Furthermore, x must fall into the 'window' [18].

3. **LVQ3**: The LVQ2.1 algorithm was based on the idea of differentially shifting the decision borders toward the Bayes limits, while no attention was paid to what might happen to the location of the m_i in the long run if this process were continued. Therefore, it seems necessary to include corrections that ensure that the m_i continue approximating the class distributions [18]. Combining these ideas, an improved algorithm can be obtained that may be called LVQ3:

$$\begin{aligned} m_i(t+1) &= m_i(t) - \alpha(t)[x(t) - m_i(t)] \\ m_j(t+1) &= m_j(t) + \alpha(t)[x(t) - m_j(t)] \end{aligned} \tag{3.79}$$

where m_i and m_j are the two closest codebook vectors to x, whereby x and m_j belong to the same class, while x and m_i belong to different classes, respectively. Furthermore, x must fall into the 'window.'

$$m_k(t+1) = m_k(t) + \epsilon\alpha(t)[x(t) - m_k(t)] \tag{3.80}$$

for $k \in i, j$, if x, m_i, and m_j belong to the same class [18].

The three options for the LVQ algorithms, namely, the LVQ1, the LVQ2.1, and the LVQ3, yield almost similar accuracies, although a different philosophy underlies each of them. The LVQ1 and the LVQ3 define a more robust process, whereby the codebook vectors assume stationary values even after extended learning periods. For the LVQ1, the learning rate can approximately be optimized for quick convergence. In the LVQ2.1, the relative distances of the codebook vectors from the class borders are optimized whereas there is no guarantee for the codebook vectors being placed optimally to describe the forms of the class borders. Therefore, the LVQ2.1 should only be used in a differential fashion, using a small value of learning rate and a relatively low number of training steps [18].

3.9 Conclusion

In this chapter, we have described the fundamental considerations of ANN which is a key component used in the work described throughout the book. We have explained the analogy of artificial neuron with the biological neuron along with a description of the commonly used activation functions. We have also described the ANN learning paradigms namely supervised and unsupervised learning and a brief note on prediction and classification using ANN. Finally, we have described the primary ANN topologies like MLP, RNN, PNN, SOM, and LVQ.

References

1. Demuth H, Beale M, Hagan M, Venkatesh R (2009) Neural network toolbox6, users guide. Available via http://filer.case.edu/pjt9/b378s10/nnet.pdf
2. Fausett LV (1993) Fundamentals of neural networks architectures, algorithms and applications, 1st edn. Pearson Education, New Delhi, India
3. Haykin S (2003) Neural networks a comprehensive foundation, 2nd edn. Pearson Education, New Delhi, India
4. Kumar S (2009) Neural networks a classroom approach. TaTa McGraw Hill, India (8th reprint)
5. Jaeger H (2002) A tutorial on training recurrent neural networks, covering BPPT, RTRL, EKF and the "echo state network" approach. GMD-Report, Fraunhofer Institute for Autonomous Intelligent Systems (AIS), 159
6. Haykins S (2002) Adaptive filter theory, 4th edn. Pearson Education, New Delhi, India
7. Goh SL, Mandic DP (2004) A complex-valued RTRL algorithm for recurrent neural networks. Neural Comput 16:2699–2713
8. Maheswari UN, Kabilan AP, Venkatesh R (2005) Speaker independent phoneme recognition using neural networks. J Theor Appl Inform Technol 12(2):230–235
9. Rao PVN, Devi TU, Kaladhar D, Sridhar GR, Rao AA (2009) A probabilistic neural network approach for protein superfamily classification. J Theor Appl Inform Technol 6(1):101–105
10. Probabilistic and general regression neural networks. Available via http://www.dtreg.com/pnn.htm
11. Jain AK, Mao J (1996) Artificial neural networks: a tutorial. IEEE Comput 29(3):31–44
12. Kohonen T (1990) The self-organizing map. Proc IEEE 78(9):1464–1480
13. Alhoneiemi E, Hollmn J, Simula O, Vesanto J (1999) Process monitoring and modeling using the self-organizing map. Integr Comput Aided Eng 6(1):3–14
14. Kaski S, Lagus K (1997) Comparing self-organizing maps. In: Proceeding of international conference on neural networks, pp 809–814
15. Bauer HU, Pawelzik K (1992) Quantifying the neighborhood preservation of self-organizing feature maps. IEEE Trans Neural Netw 3(4):570–579
16. Kohonen T, Kaski S, Lagus K, Salojarvi J, Honkela J, Paatero V, Saarela A (2000) Self organization of a massive document collection. IEEE Trans Neural Netw 11(3):574–585
17. Bullinaria JA (2000) A learning vector quantization algorithm for probabilistic models. Proc EUSIPCO 2:721–724
18. Kohonen T, Hynninen J, Kangas J, Laaksonen J, Torkkola K (1995) LVQ PAK, the learning vector quantization program package, LVQ Programming Team of the Helsinki University of Technology, Laboratory of Computer and Information Science, Version 3.1, Finland

Chapter 4
Sounds of Assamese Language

Abstract This chapter intends to provide a brief outline about the sounds of Assamese language as well as some specific phonetic features. A brief note on dialects of Assamese language is also provided. The description included in this chapter is mostly from certain background study confined to some standard books and some visual inspection of Assamese sounds done using certain speech analysis softwares.

Keywords Assamese · Phoneme · Vowel · Consonant · Place of articulation · Manner of articulation · Dialects

4.1 Introduction

Production of speech signal is a physical event that occurs due to the utterance of sounds belonging to a language. Speech signal is non-stationary, mostly varying within a few milliseconds, because of the changing characteristics of the articulator involved in the production mechanism. Speech signal can be analyzed as a sequence of sounds which share some common acoustic and articulatory properties for a short duration of time. For each sounds, there is some particular position of the vocal folds, tongue, lips, teeth, velum, and jaw, which are the vocal tract articulators involved in the production process [1–3]. However, due to the different history of natural evolution of any language, the basic sound units present are different from language to language [4]. Depending on the nature of excitation source, the sound produced can be divided into two broad classes—voiced and unvoiced [1, 2, 4]. These sounds can be analyzed from articulatory points of view also. It relates linguistics features of sounds to positions and movements of the articulators [1]. A phoneme which is the smallest meaningful unit in the phonology of a language is a linguistic abstraction in the cognitive level. The physical sound produced when a phoneme is articulated is called a phone. Infinite number of phones may correspond to the same phoneme since repeated pronunciations of the phoneme differ from one another for the same

M. Sarma and K. K. Sarma, *Phoneme-Based Speech Segmentation Using Hybrid Soft Computing Framework*, Studies in Computational Intelligence 550,
DOI: 10.1007/978-81-322-1862-3_4, © Springer India 2014

speaker itself [1]. Every language has a small set of 20–40 phonemes, which is sufficient to represent the sounds present in the language. In general, the phonemes of a language are divided into two broad classes called vowel phonemes and consonant phonemes. They are further classified in terms of manner of articulation (MOA) and place of articulation (POA). MOA concerns how the vocal tract restricts airflow. If airflow is completely stopped by an occlusion, it creates a stop consonant; vocal tract constriction in varying degree occurs in fricatives and vowels, whereas lowering the velum causes nasal sounds. Similarly, POA refers to the location in the vocal tract in the upper wall [1, 2, 4]. Thus, vowels can be classified as front, mid, and back vowels as well as diphthongs, whereas consonants are classified as stop consonants, nasals, fricatives, affricates, and semivowels.

Assamese is a major language spoken in the north-eastern part of India. It is the official language of the state of Assam, pronounced as ɑxɑmija by the native speakers. Assamese was accorded the status of one of the official language of the state along with English by the state's Official Language Act of 1960. It is considered as the lingua franca of the entire Northeast India. The Assamese language is the eastern most member of the Indo-European family tree. The history of Assamese can be traced back to very early times. The Assamese language grew out of Sanskrit, the ancient language of the Indian subcontinent. However, its vocabulary, phonology, and grammar have substantially been influenced by the original inhabitants of Assam such as the Bodos and the Kacharis. Assamese developed from Magadhi Prakrit, the eastern branch of Apabhramas that followed Prakrit. Assamese has been a borrowing language, and during its long history of evolution, it has enriched its vocabulary by acquisition from all the non-Aryan languages [5]. A brief introduction of the sounds of Assamese language as well as some specific phonetic features is described in the following sections.

4.2 Formation of Assamese Language

Assamese has originated from the old Indo-Aryan dialects, but the exact nature of its origin and growth is not clear yet [5–7]. However, it is supposed that Assamese evolved from a *Magadhi Prakrit*. According to linguist Suniti Kumar Chatterji, the *Magadhi Prakrit* in the east gave rise to four Apabhramsa dialects: *Radha*, *Vanga*, *Varendra*, and *Kamarupa*, and the Kamarupa Apabhramsa spread to the east and north of the Ganges gave rise to the North Bengal dialects in West Bengal and Assamese in the Brahmaputra Valley. Though early compositions in Assamese exist from the thirteenth century, the earliest relics of the language can be found in paleographic records of the *Kamarupa* kingdom from the fifth to the twelfth century. Assamese language features have been discovered in the ninth century Charyapada, which are Buddhist verses discovered in 1907 in Nepal. These came from the end of the Apabhramsa period. Early compositions matured in the fourteenth century, during the reign of the Kamata king Durlabhnarayana of the Khen dynasty, when Madhav Kandali composed the Kotha Ramayana. Since the time of the Charyapada,

Assamese has been influenced by the languages belonging to the Sino-Tibetan and Austroasiatic families. Assamese became the court language in the Ahom kingdom by the seventeenth century [5–8].

4.3 Phonemes of Assamese Language

Speech sounds can be basically classified as voiced and unvoiced. This classification is based on the source which excites the vocal tract filter. If the vocal tract filter is excited by a source signal with some periodic vibration, then sound unit is called voiced sounds. On the other hand, if the excitation source signal is noise like, then the resulting sound unit is called unvoiced [3]. But this classification is not enough to describe the variations of sound units identified in a language. Another popular classifications of the sounds of a language are vowels and consonants. Vowels are the signal having high energy and low frequency. But consonants are gestures of lips and tongue with relatively low energy [9], and therefore, a vowel is always required to carry the message or information contained in the consonant sound. Sounds are best described by classifying them according to the POA and MOA. MOA concerns how the vocal tract restricts airflow. Completely stopping airflow by an occlusion creates a stop consonant. Vocal tract constrictions of varying degree occur in fricatives and vowel, whereas lowering the velum causes nasal sounds [1–3, 9]. On the other hand, POA provides some finer discrimination of phonemes [1]. It refers to the point of narrowest vocal tract constriction. Languages differ considerably depending on which places of articulation are used within the various MOA. Vowel, stops, nasal, and fricatives are present in almost every language, what brings distinction among the languages is the POA. Further, some phonemes are viewed as having phoneme subsequences. Two classes of such phonemes are there—diphthongs and affricates. Diphthong produces when vowel is followed by a high vowel, and affricate produces when stop is followed by a fricative [1, 3]. The places of articulation are mostly associated with consonants. This is because consonants use narrow constriction unlike vowels. Eight places of articulation are observed along the vocal tract, and accordingly, the consonants are called labials, dental, alveolar, palatal, velar, uvular, pharyngeal, and glottal. If both lips constricts, the sound is called bilabial, while if lower lip contacts the upper teeth, then the sound is labiodental. When the tongue tip touches the edge or back of the upper teeth, the sound produced is called dental, and when the tongue tip approaches or touches the alveolar ridge, the sound produced is called alveolar. When the tongue blade or dorsum constricts with the hard palate, the sound is called palatal, and if the tongue tip curls up, the sound is retroflex. When the dorsum approaches the soft palate, the sound is called velar, and sound is called uvular when the dorsum approaches the uvula. The pharynx constricts to produce the pharyngeal sounds, and the vocal tracts either close or constrict to produce the glottal stops [1–3, 9]. This way the POA brings variation among the sound units as well as the belonging languages. The following sections describe the sounds present in Assamese language.

4.3.1 Vowels

Vowels are the phonemes having highest energy and range in duration between 50 and 400 ms in normal speech [1]. Vowels are produced by exciting a fixed vocal tract with quasi-periodic pulses of air caused by vibration of the vocal cords. Due to the excitation, the cross sectional area of the vocal tract varies producing resonant frequencies and the manner of this variation determines the vowel produced. The variation in the vocal tract shape is primarily influenced by the positions of the jaw, lips, and velum [3]. For example, in forming the vowel /a/ as in the Assamese word 'pat' meaning leaf in English, the vocal tract is open in the front and constricted at the back by the tongue; on the other hand, the vowel /i/ as in the Assamese word 'bil' meaning small lake in English is formed by raising the tongue toward the palate, thus causing a constriction at the front, but increasing the opening at the back of the vocal tract. Thus, from signal processing point of view, it can be stated that different vowels are produced due to the different vocal tract filter transfer function excited by the same excitation source. The resonance of the vocal tract filter is popularly known as formant frequencies. Formants are measured as an amplitude peak in the frequency spectrum of the vowel sound using a spectrogram [9]. Therefore, vowels can be distinguished by its first and second formant frequencies. According to the relative frequency of the first formant, F1, vowels are classified as low vowels and high vowels. The higher the F1 value, the lower the vowel. This is called vowel height, which is inversely correlated to F1. As the tongue body is raised to decrease the cross sectional area of the constriction formed between the dorsum of the tongue and the palate, F1 decreases in frequency in case of high vowels [2]. International phonetic alphabet (IPA) [10] also defines close vowels and open vowels which are related to the opening and closing of the jaw, respectively. Similarly, according to relative frequency of second formant, F2, on the spectrum, vowels are classified as front, mid, and back vowels. It depends on the front–back position of the tongue body. Since the tongue body is displaced forward while maintaining a narrowing in the lower pharynx, the F2 increases to a maximum value in case of front vowels. Hence, the higher the F2, the fronter the vowel. But in case of back vowels, the tongue body displaced to backed positions, which lead to proper condition of the constriction to give a low F2 value [2]. This is known as vowel backness. The maximum value of F2 is higher for the high vowels than for the low vowels [2, 9]. IPA identifies five classes of vowel backness as front, near front, central, near back, and back vowels [10].

Every language has a specific set of vowels, each of which contrast with one another so that they make different words. Many world languages such as Spanish, Japanese use just five vowels called /a/, /e/, /i/, /o/, and /u/. As identified by G. C. Goswami, J. Tamuli etc., Assamese language has eight vowels /i/, /e/, /ɛ/, /a/, /ɒ/, /ɔ/, /o/, and /u/ [5, 7]. Out of these, /i/, /e/, and /ɛ/ are front vowels and /ɒ/, /ɔ/, /o/, and /u/ are back vowels, whereas /a/ is the only central vowel. Figure 4.1 shows the vowels of the language according to the height and backness of the tongue body. Some example Assamese words containing these eight vowels are given in Table 4.1. The acoustic waveform of the eight Assamese vowels are shown in

	Front	Central	Back
High	i		u
Higher mid	e		o
Lower mid	ɛ		ɔ
Low		a	ɒ

Fig. 4.1 Assamese vowel phonemes [5]

Table 4.1 Assamese words containing the eight vowels

Vowel	Word	Meaning
/i/	/sin/ presents eight-way contrast	Spot
	/bfiinifii/ presents six-way contrast	Brother-in-law
	/ritu/ presents five-way contrast	Season
/e/	/bes/ presents eight-way contrast	Fine
	/abeli/ presents six-way contrast	Afternoon
	/renu/ presents five-way contrast	Pollen
/ɛ/	/bɛl/ presents eight-way contrast	A kind of fruit
/a/	/bas/ presents eight-way contrast	To choose
	/mɒdafii/ presents six-way contrast	Drunkard
	/ati/ presents five-way contrast	Bundle
/ɒ/	/bɒl/ presents eight-way contrast	Strength
/ɔ/	/bɔl/ presents eight-way contrast	Lets go
	/rɔdali/ presents six-way contrast	Sunny
	/pɔtu/ presents five-way contrast	Expert
/o/	/bol/ presents eight-way contrast	Color
	/bfiogi / presents six-way contrast	One who enjoys
/u/	/bul/ presents eight-way contrast	To walk
	/bfiugi/ presents six-way contrast	Having suffered
	/uti/ presents five-way contrast	Having floated

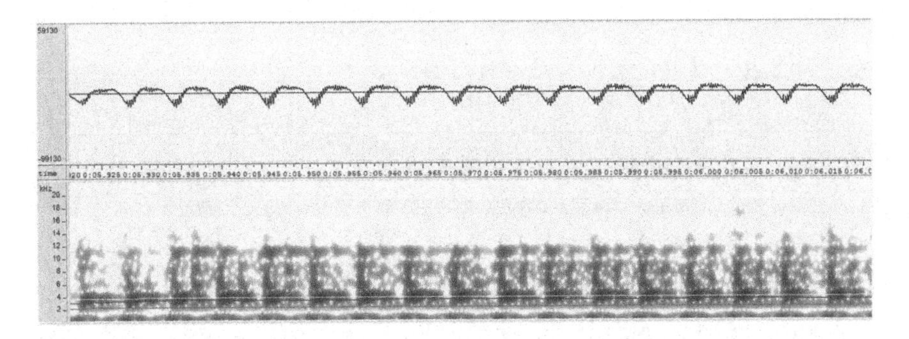

Fig. 4.2 Speech waveform and its spectrum for vowel /i/ extracted from word 'pit' meaning act of biting

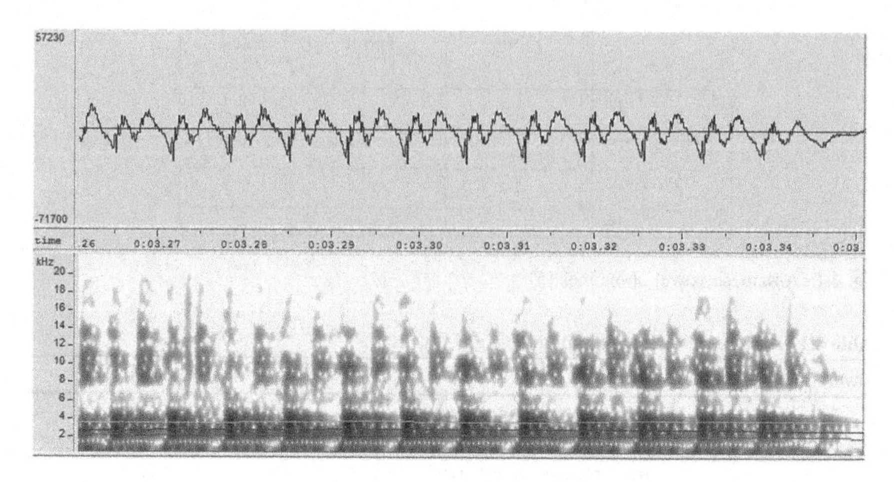

Fig. 4.3 Speech waveform and its spectrum for vowel /e/ extracted from word 'petu' meaning intestines

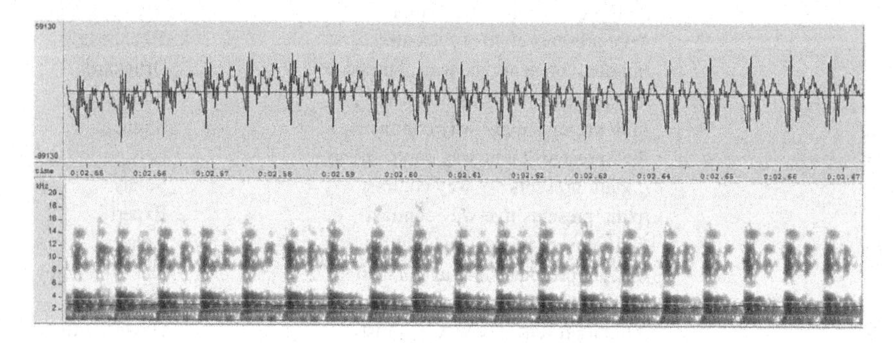

Fig. 4.4 Speech waveform and its spectrum for vowel /ɛ/ extracted from word 'pɛt' meaning stomach

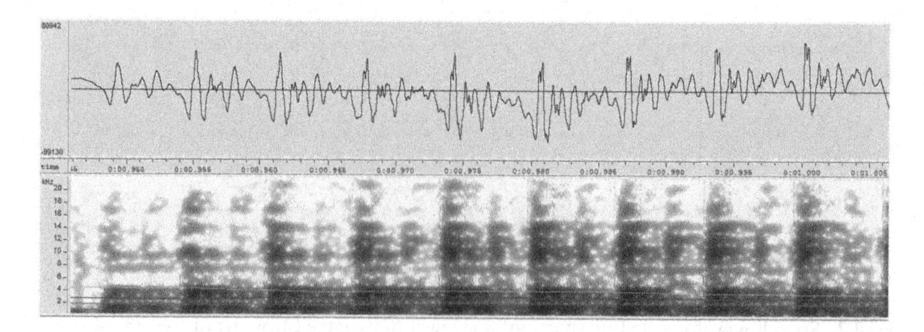

Fig. 4.5 Speech waveform and its spectrum for vowel /a/ extracted from word 'pat' meaning leaf

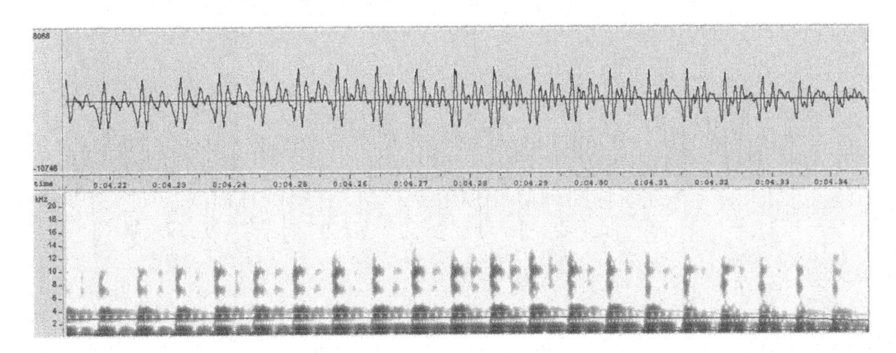

Fig. 4.6 Speech waveform and its spectrum for vowel /ɒ/ extracted from word 'pɒt' meaning scenario

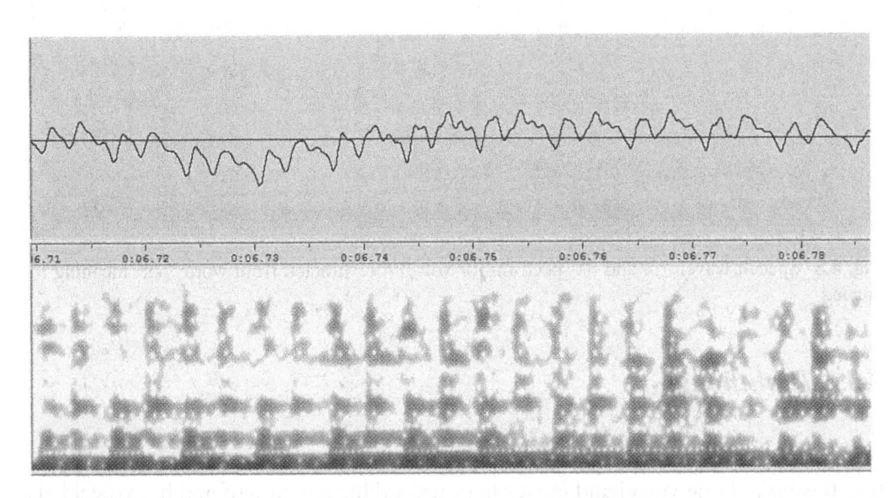

Fig. 4.7 Speech waveform and its spectrum for vowel /ɔ/ extracted from word 'pɔti' meaning husband

Figs. 4.2, 4.3, 4.4, 4.5, 4.6, 4.7, 4.8, and 4.9 along with their spectrum obtained from spectrogram. These eight vowels present three different types of contrasts [5]. Firstly, eight-way contrast in closed syllables, and in open syllables when /i u/ do not follow in the next immediate syllable with intervention of a single consonants except the nasal. Again, it shows six-way contrast in open syllables with /i/ occurring in the immediately following syllable with intervention of any single consonant except the nasals, or except with nasalization, and finally, five-way contrast in open syllables when /u/ occurs in the immediately following syllable with a single consonant intervening [5].

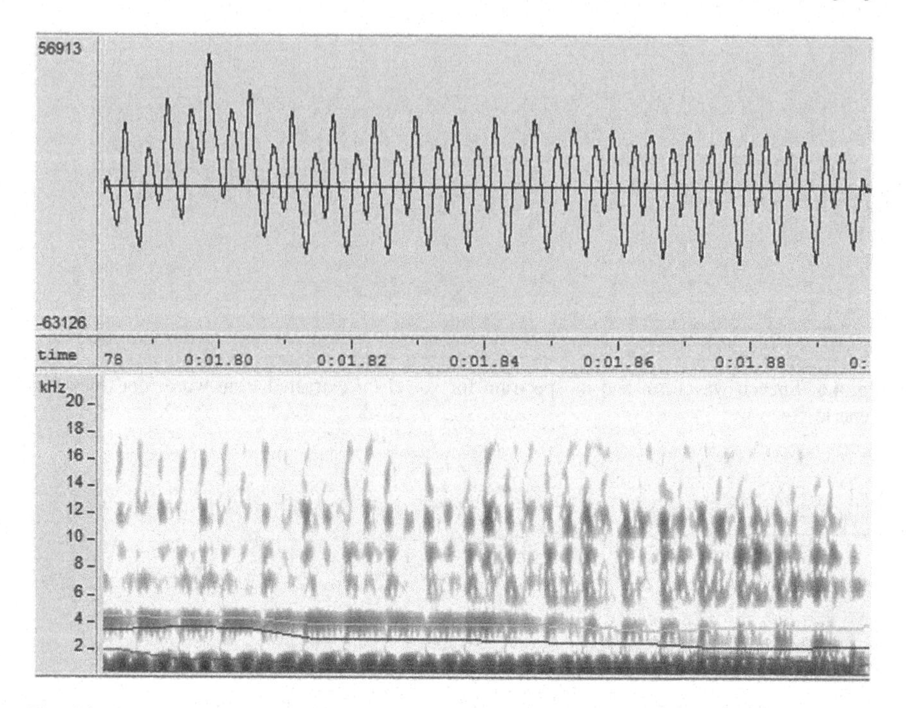

Fig. 4.8 Speech waveform and its spectrum for vowel /o/ extracted from word 'pot' meaning to burried

4.3.2 Diphthongs

A diphthong is a gliding monosyllabic speech item that starts at or near the articulatory position of one vowel and moves to or toward the position of another vowel [3]. The phonemic inventory of Assamese is made up of fifteen diphthongs where two are nasalized. These are /ei/, /iu/, /ia/, /ui/, /ou/, /ua/, /oi/, /eu/, /ɛa/, /ɔi/, /ɔu/, /ai/, and /au/ [11].

4.3.3 Stop Consonant

Stops or plosive are speech sounds related to the time interval when the vocal tract has a narrow constriction and the immediately adjacent points in time when there may be an implosion or a release of the articulatory structures from this constricted configuration. Due to a sequence of articulatory and acoustic events, a closure is made at a particular point along the vocal tract which is followed by a time interval when the vocal tract remains closed and after that the closure is released. When vocal tract moves toward or away from the constricted configuration, the perceptual correlates

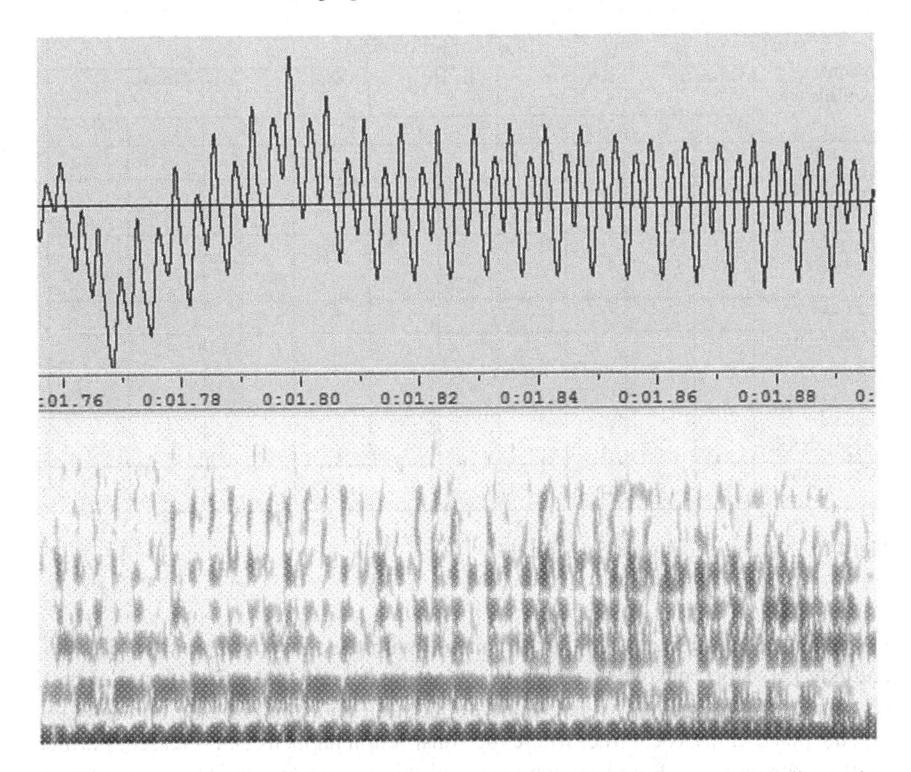

Fig. 4.9 Speech waveform and its spectrum for vowel /u/ extracted from word 'puti' meaning burried

for a consonant features are obtained which results in the stop consonants [2]. The stops may be voiced or voiceless depending on whether the vocal cords are able to vibrate or not during the total closure of the vocal tract [3]. On the other hand, the stops may be aspirated or unaspirated. In some stop consonants, after releasing the closure and before the following vowel starts, there is a small delay in which the air rushes out producing some aspiration. The resulting stop consonant produced is called aspirated stop. Unaspirated stops have low periodic energy at low frequency in the first 20 ms after the closure is released. The interval between the release of a stop and the start of a following vowel is called the voice onset time (VOT). For aspirated stops, the VOT is long in comparison with the unaspirated stops [9]. Languages such as French, Spanish allows voiceless stops with no aspiration and fully voiced stops. But languages such as English, German have voiced stops contrast with voiceless stops [9]. Assamese is a language where both voiced and unvoiced aspirated stop consonants and voiced and unvoiced unaspirated stop consonants are identified [5]. The voiced unaspirated stops of Assamese language are /b/, /d/, and /g/, all of which has the aspirated version /bɦ/, /dɦ/, and /gɦ/. Similarly, Assamese

Manner of Articulation	Places of Articulation									
	Bilabial		Alveolar		Palatal		Velar		Glottal	
	Vl	Vd	Vl	Vd	Vl	Vd	Vl	Vd	Vl	Vd
Nasal		m		n				ŋ		
Unaspirated Stop	p	b	t	d			k	g		
Aspirated Stop	pɦ	bɦ	tɦ	dɦ			kɦ	gɦ		
Fricative			s	z			x			ɦ
Approximant				ɹ		j				
Tap				ɾ						
Lateral Approximant				l						
Continuants		w								

Fig. 4.10 Assamese consonant phonemes [5]

has unvoiced unaspirated stops /p/, /t/, and /k/ with their aspirated version /pɦ/, /tɦ/, and /kɦ/ [5]. Stop consonants are further classified as bilabial, labio-dental, dental, alveolar, retroflex, palatal, velar, uvular, epiglottal, and glottal based on the POA, i.e., the place of the vocal tract where the constriction happens and released. Further glottal stops are also found, which closes glottis for 30–60 ms [1]. Among these, in Assamese language, bilabial (/p/, /b/), alveolar (/t/, /d/), and velar (/k/, /g/) stop consonants are observed [5]. For /b/, constriction is at the lips; for /d/, constriction is back of the teeth; and for /g/, constriction is near the velum. Figure 4.10 shows the consonants of the language. Figures 4.11, 4.12, 4.13, 4.14, 4.15, and 4.16 show the speech waveform of the stop consonants /b/, /t/, /d/, and /k/ observed in some Assamese words given in Table 4.2.

4.3.4 Nasals

The nasal phonemes are produced when glottal excitation and the vocal tract totally constrained at some point along the oral passage way, and therefore, the velum is lowered so that the air flows through the nasal tract, with sound being radiated at the nostrils [1, 3]. The nasal consonants /m/, /n/, and /ŋ/ are typically observed in Assamese language [5]. During the production of nasal sounds, the oral cavity although constricted toward the front, it is still acoustically coupled to the pharynx, and hence, the mouth serves as a resonant cavity that traps acoustic energy at certain natural frequencies. The three nasal consonants are distinguished by the POA that means the place along the oral cavity at which total constriction is happened. For /m/, constriction is at the lips; for /n/, constriction is at the back of the teeth; and

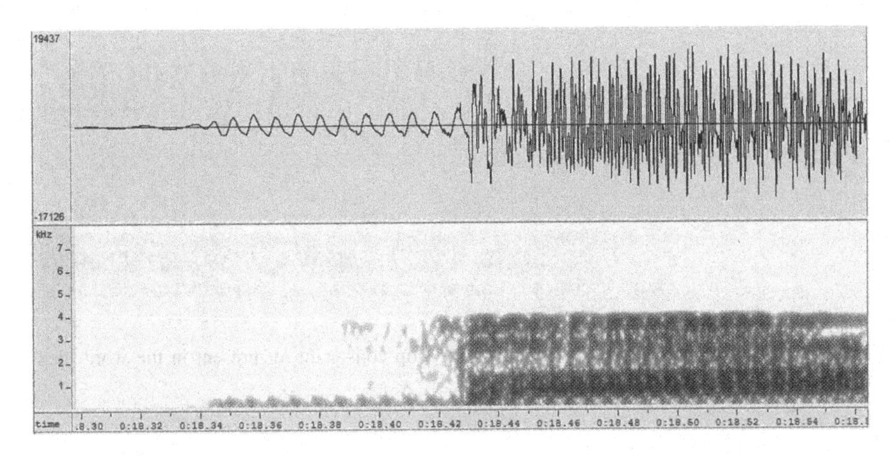

Fig. 4.11 Speech waveform and its spectrum for stop consonant /b/ present in the word 'bagh' meaning tiger

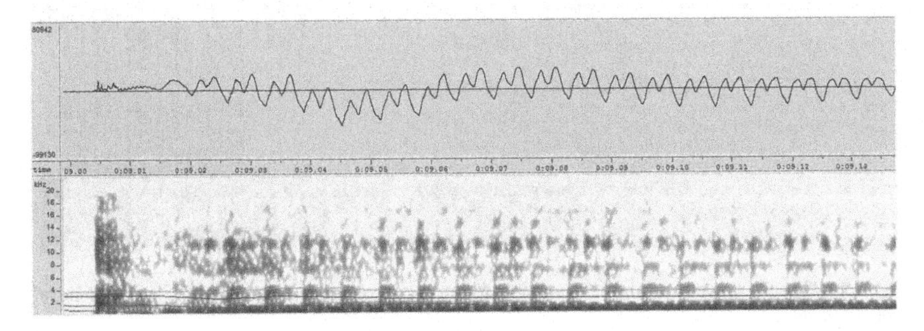

Fig. 4.12 Speech waveform and its spectrum for stop consonant /t/ present in the word 'top' meaning drop

for /ŋ/, the constriction is just forward of the velum. Hence, they are called bilabial, alveolar, and velar nasal, respectively. Some example Assamese words are 'mɑn' meaning mind, 'nat' meaning drama, and rɑŋ meaning color and joy. Nasal sounds have waveforms similar to that of vowel sounds since both have periodic excitation source. But the nasal sounds have significantly week energy because of the sound attenuation in the nasal cavity [1]. When a vowel preceded a nasal consonant, the velum often lowered due to which the nasalization is observed in the vowel. Due to nasalization, an additional nasal resonance is observed near F1 in the spectrum of the vowel. In Assamese language also vowel nasalization is observed in the words like

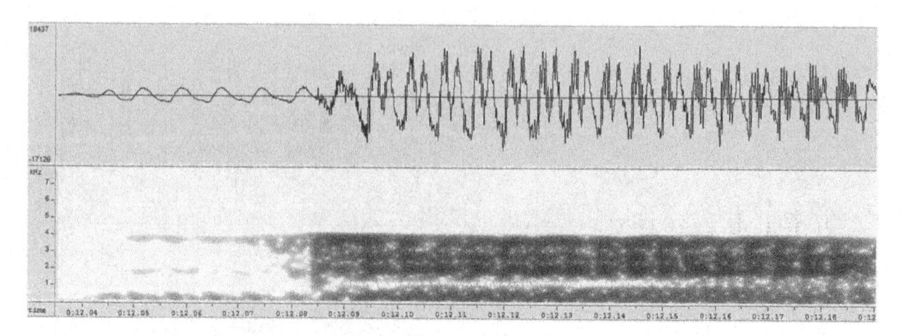

Fig. 4.13 Speech waveform and its spectrum for stop consonant /d/ present in the word 'dex' meaning nation

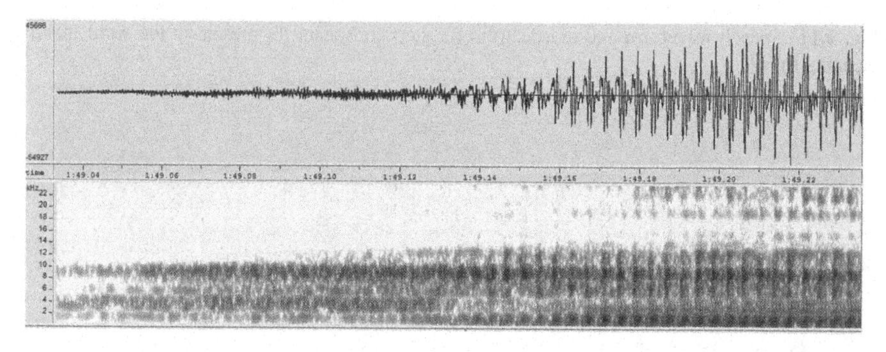

Fig. 4.14 Speech waveform and its spectrum for stop consonant /k/ present in the word 'kendrɑ' meaning center

4.3.5 Fricatives

Fricatives are the sounds produced due to the constriction in the oral tract or pharynx or glottis. Unvoiced fricatives are produced by exciting the vocal tract by a steady air flow which becomes turbulent in the region of a constriction in the vocal tract. However, voiced fricatives have two sources of excitation. One excitation source is at the glottis since the vocal cords are vibrating to produce the voiced fricative. The other excitation source is the turbulent air flow since the vocal tract is constricted at some point forward of the glottis. The location of the constriction determines which fricative sound is produced. In Assamese language, voiceless alveolar fricative /s/ and velar fricative /x/ are observed. Further, voiced alveolar fricative /z/ and voiced glottal fricative /ɦ/ are also identified. Presence of unvoice velar fricative /x/ is an unique property of the language unlike other Indian language. Figures 4.15 and 4.16

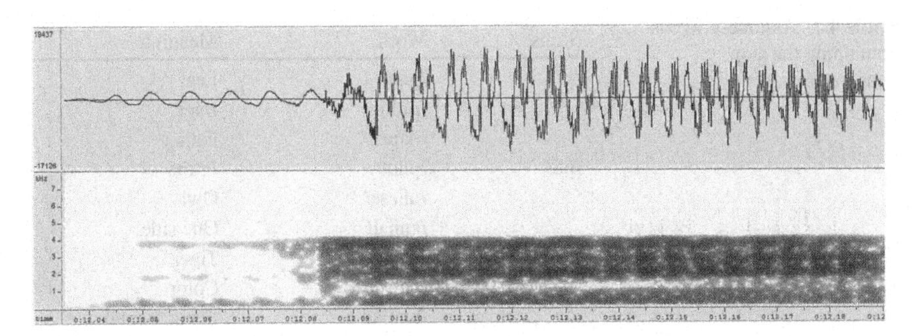

Fig. 4.15 Speech waveform and its spectrum for fricative /x/ present in the word 'Axa' meaning hope

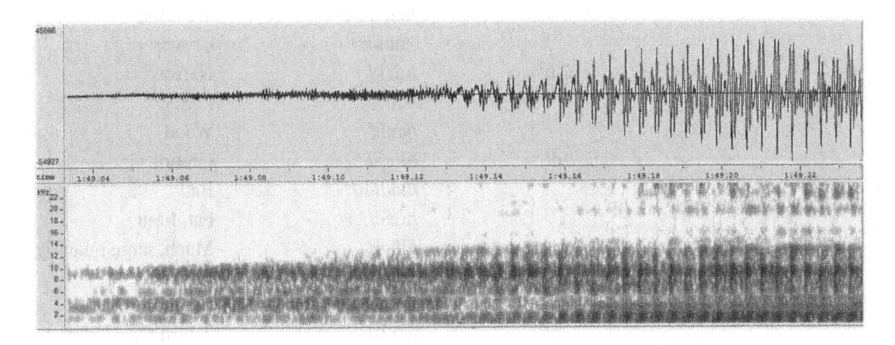

Fig. 4.16 Speech waveform and its spectrum for fricative /h/present in the word 'ɦaɦi' meaning asking to lough

show the speech waveform of the sound /x/ and /ɦ/. The non-periodic nature of excitation is observed in the waveform of fricative /x/, whereas in the waveform of /ɦ/, the presence of two excitation source is readily observed.

4.3.6 Affricates

Affricates start like a stop but release as a fricative. Recent studies like Sarma et al. [12] has reported the presence affricates /ts/ and /dz/ in Assamese language in the words like 'χoitsɑ' meaning crop and 'dzɑrna' meaning small waterfall.

Table 4.2 Assamese words containing the stop consonants

Stops	Word	Meaning
/p/	/pat/	Leaf
	/pithi/	Back
	/bapɛk/	Father
/pɦ/	/pɦɑl/	Fruits
	/pɦɛsa/	Owl
	/ɛpɦal/	One side
/b/	/bagɦ/	Tiger
	/bɑrɑn/	Color
	/buniadi/	Fundamental
/bɦ/	/bɦitɑr/	Interior
	/bibɦuti/	Wealth
	/ɑxɑmbɦab/	Impossible
/t/	/top/	Drop
	/batɔrua/	Travellar
	/mukut/	Crown
/tɦ/	/tɦɛk/	Narrow
	/sitɦi/	Letter
	/katɦ/	Wood
/d/	/dɑra/	Groom
	/baduli/	Bat
	/mɛd/	Fat, limit
/dɦ/	/dɦer/	Much, large quantity
	/adɦunik/	Modern
	/nixedɦ/	Prohibition
/k/	/Kuli/	A bird Cuckoo
	/akɦax/	The sky
	/xɑmazik/	Dream
/kɦ/	/kɦel/	Game
	/akɦjan/	Story, Fable
	/nɑkɦ/	Neil
/g/	/git/	Song
	/bagisa/	Garden
	/xɑɦɑzog/	Cooperation
/gɦ/	/gɦati/	Deficit
	/ɑgɦat/	Injury
	/magɦ/	An Assamese month

4.3.7 Semi Vowels

Semivowels are special type of MOA which has vowel like nature. They are similar to high vowels but characterized by a gliding transition in the vocal tract area function between adjacent phonemes, and hence, they are influenced by contextual dependency. The semivowels are close in articulation to a high vowel. For example, semivowel /j/ is similar to high vowel /i/ and /w/ is similar to /u/. Assamese language has the semivowels like /l/, /r/, /j/, and /w/ in the words like 'mukul' meaning bud, 'rɔd' meaning sunshine, 'ɑnjai' meaning injustice, and 'swami' meaning husband. The voiced bilabial frictionless continuants /w/ and voiced palatal frictionless continuants /j/ have been treated as a rare phoneme of the Assamese language because they are rarely distributed. /j/ is the only palatal phoneme of the language.

4.4 Some Specific Phonemical Features of Assamese Language

Retaining certain features of its parent Indo-European family, Assamese has got many unique phonological characteristics. There is several phonological uniqueness of Assamese sounds which shows minor variations when spoken by people of different dialects of the language. This makes Assamese speech unique, and hence, the language is a rich repository for study by speech and language researchers. Following here are few special features of the phonetic inventory of the language.

1. **Alveolar Stops**: The Assamese phoneme inventory is unique in the Indic group of languages in its lack of a dental-retroflex distinction in coronal stops. Historically, the dental stops and retroflex stops both merged into alveolar stops. This makes Assamese resemble non-Indic languages in its use of the coronal major POA.
2. **Voiceless Velar Fricative**: Unlike most eastern Indic languages, Assamese is also noted for the presence of the voiceless velar fricative /x/ historically derived from what used to be coronal sibilants. The derivation of the velar fricative from the coronal sibilant [s] is evident in the name of the language in Assamese; some Assamese prefer to write ɑxɑmija instead of *Asomiya/Asamiya* to reflect the sound, represented by [x] in the IPA. This sound appeared in the phonology of Assamese as a result of lenition of the three Sanskrit sibilants. It is present in other nearby languages, like Chittagonian.
   The sound is variously transcribed in the IPA as a voiceless velar fricative /x/, a voiceless uvular fricative /χ/, and a voiceless velar approximant by leading phonologists and phoneticians. Some variations of the sound is expected within different population groups and dialects, and depending on the speaker, and quality of recording, all three symbols may approximate the acoustic reading of the actual Assamese phoneme [13].
3. **Velar Nasal**: Assamese and Bengali, in contrast to other Indo-Aryan languages, use the velar nasal /η/ extensively. In many languages, the velar nasal is always attached to a homorganic sound, whereas in Assamese, it can occur intervocalically [11].

4.5 Dialects of Assamese Language

Along with distinctive phonemical diversity in Assamese language, several regional dialects are typically recognized in different regions of the state. Banikanta Kakati identified two dialects which he named *Eastern and Western dialects* [5]. However, recent linguistic studies have identified four dialect groups, listed below from east to west [11]:

- *Eastern group* spoken in and other districts around Sibsagar district.
- *Central group* spoken in present Nagaon district and adjoining areas.
- *Kamrupi group* spoken in undivided Kamrup, Nalbari, Barpeta, Darrang, Kokrajhar, and Bongaigaon districts.
- *Goalparia group* spoken primarily in the Dhubri and Goalpara districts and in certain areas of Kokrajhar and Bongaigoan districts.

Dialects vary primarily with respect to phonology, most of the time due to the variation in the occurrence of the vowel sounds. As for example, the Assamese word representing the vegetable *gourd* is pronounced differently in the above dialects. In Kamrupi dialect, this is pronounced as /kumra/ but in standard Assamese as /komora/. Similarly, in standard Assamese, *king* is pronounced as /rɑza/, whereas in Kamrupi and Goalparia dialect, it is pronounced as /raza/ [6]. Thus, region to region and speaker to speaker the language shows some notable variations.

4.6 Conclusion

A brief introduction of the sounds present in Assamese language is described in this chapter. A brief note on dialects of Assamese language is also provided. Presence of various unique sounds in Assamese language reflects the importance of the language in speech and natural language processing based applications.

References

1. Shaughnessy O' D (2000) Speech communications, human and machine douglas. IEEE Press, New york
2. Stevens KN (2000) Acoustic phonetics. 1st MIT Press paperback edn. The MIT Press, Cambridge, Massachusetts and London, England
3. Rabiner LR, Schafer RW (2009) Digital processing of speech signals. Pearson Education, Dorling Kindersley (India) Pvt. Ltd, Delhi
4. Bhaskararao P (2011) Salient phonetic features of Indian languages in speech technology. Sadhana 36(5):587–599
5. Goswami GC (1982) Structure of Assamese, 1st edn. Department of publication, Gauhati University, Guwahati
6. Goswami UN (1978) An introduction to Assamese. Mani-Manik Prakash, Guwahati

7. Goswami GC, Tamuli JP (2003) Asamiya. The Indo-Aryan languages. Routledge, London, pp 391–443
8. Guha A (1983) The Ahom political sysem. Soc Sci 11(12):334
9. Ladefoged P, Disner SF (2012) Vowels and consonants, 3rd edn. Wiley-Blackwell Publishing Ltd, West Sussex
10. International Phonetic Alphabet. Available via http://www.langsci.ucl.ac.uk/ipa/ipachart.html
11. Resource Center for Indian Language Technology Solutions, IIT Guwahati. Available via http://www.iitg.ernet.in/rcilts/phaseI/languages/asamiya.htm
12. Sarma BD, Sarma M, Sarma M, Prasanna SRM (2013) Development of Assamese phonetic engine: eome issues. In: Proceedings of INDICON 2013, IIT Bombay, Mumbai, India
13. The Assam Tribune Editorial (2006) The X sound in Assamese, language, Mar 5

Chapter 5
State of Research of Speech Recognition

Abstract In this chapter, a brief overview derived out of detailed survey of speech recognition works reported from different groups all over the globe in the last two decades is described. Robustness of speech recognition systems toward language variation is the recent trend of research in speech recognition technology. To develop a system which can communicate with human being in any language like any other human being is the foremost requirement of any speech recognition technology for one and all. From the beginning of commercial availability of the speech recognition system, the technology has been dominated by the Hidden Markov Model (HMM) methodology due to its capability of modeling temporal structures of speech and encoding them as a sequence of spectral vectors. However, from the last 10 to 15 years, after the acceptance of neurocomputing as an alternative to HMM, ANN-based methodologies have started to receive attention for application in speech recognition. This is a trend worldwide as part of which a few works have also reported by researchers. India is a country which has vast linguistic variations among its billion plus population. Therefore, it provides a sound area of research toward language-specific speech recognition technology. This review also covers a study on speech recognition works done specifically in certain Indian languages. Most of the work done in Indian languages also uses HMM technology. However, ANN technology is also adopted by a few Indian researchers.

Keywords Automatic speech recognition · Hidden Markov model · Artificial neural network · Indian language

5.1 Introduction

The problem of automatic speech recognition (ASR) was at the forefront of research till 1930 when the first electronic voice synthesizer was designed by Homer Dudley of Bell Laboratories. After that, ASR lost its fascination among the speech processing

M. Sarma and K. K. Sarma, *Phoneme-Based Speech Segmentation Using Hybrid Soft Computing Framework*, Studies in Computational Intelligence 550,
DOI: 10.1007/978-81-322-1862-3_5,

community. Probably, that was the starting of research in the direction of designing a machine that can mimic the human capability of speaking naturally and responding to spoken languages. Initial developments covers simple machine that responds to isolated sounds. Recently, speech recognition technology has risen to such a height that a large community of people now talk to their mobile smart phones, asking them to send e-mail and text messages, search for directions or find information on the Web. However, speech recognition technology is still far from having a machine that converses with humans on any topic like another human. In the present times, research in speech recognition concentrates on developing systems that can show robustness for variability in environment, speaker, and language. India is a linguistically rich country having 22 official languages and hundreds of other sublanguages and dialects spoken by various communities covering the billion plus population. Communication among human beings is dominated by spoken language. Therefore, it is natural for people to expect speech interfaces with computer, which can speak and recognize speech in native language. But speech recognition technology in the Indian scenario is restricted to small amount of people who are both computer literate and proficient in written and spoken English. In this domain, extensive researches are going on all over India among various groups to make appropriate ASR systems in specific languages.

Initial speech recognition systems were related to isolated word recognition designed to perform special task. But in the last 25 years, certain dramatic progress in statistical methods for recognizing speech signals has been noticed. The statistical approach makes use of the four basic principles, which are Bayes decision rule for minimum error rate, probabilistic models, e.g., hidden Markov models (HMMs) or conditional random fields (CRF) for handling strings of observations like acoustic vectors for ASR and written words for language translation, training criteria and algorithms for estimating the free model parameters from large amounts of data and the generation or search process that provides the recognition or translation result [1, 2]. The speech recognition research is dominated by the statistical approaches specifically by the HMM technology till the last one decade. It is the improvement provided by HMM technology for speech recognition in the late 1970s and simultaneous improvement in speed of computer technology, due to which the ASR systems have become commercially viable in the 1990s [3–5]. But recently in the last decade, all over the world, the ANN-based technologies are receiving greater attention. This is due to the fact that ANN models are composed of many nonlinear computational elements operating in parallel and arranged in the pattern of biological neural network. It is expected that human neural network like models may ultimately be able to solve the complexities of speech recognition system and provide human-like performance. This is possible by using ANN-based approaches.

In this chapter, we have highlighted some works related to the research and development of ASR technology during the last decade so as to provide a picture of the fundamental progress that has been made in the large variety of world languages. Initially, a glance of early speech recognition technologies developed among various world languages are included.

5.2 A Brief Overview of Speech Recognition Technology

Early speech recognition systems used the acoustic–phonetic theories of speech to determine the feature [1]. Due to the complexity of human language, the inventors and engineers first focused on number or digit recognition. The first speech recognition system was built in Bell Laboratories by Davis et al. [6] in 1952, which could understand only isolated digits for a single speaker. They used the formant frequency measured during vowel regions of each digit as a feature. During 1950–1970, laboratories in the USA, Japan, England, and the former Soviet Union developed other hardware dedicated to recognizing spoken sounds, expanding speech recognition technology to support four vowels and nine consonants [7–13]. In the 1960s, several Japanese laboratories demonstrated their capability of building special purpose hardware to perform a speech recognition task. Among them, vowel recognizer of Suzuki and Nakata [7] at the Radio Research Laboratory in Tokyo, the phoneme recognizer of Sakai and Doshita [8] at Kyoto University, and the digit recognizer of NEC Laboratories [9] were most notable. The work of Sakai and Doshita involved the first use of a speech segmenter for analysis and recognition of speech in different portions of the input utterance. An alternative to the use of a speech segmenter was the concept of adopting a non-uniform time scale for aligning speech patterns [11, 12], dynamic programming for time alignment between two utterances known as dynamic time warping, in speech pattern matching [12] etc. Another milestone of the 1960s is the formulation of fundamental concepts of linear predictive coding (LPC) [14, 15] by Atal and Itakura, which greatly simplified the estimation of the vocal tract response from speech waveforms. Development during the 1970s includes the first speech recognition commercial company called Threshold Technology founded by Tom Martin [1] and Speech Understanding Research (SUR) program founded by Advanced Research Projects Agency (ARPA) of the US Department of Defense [1]. Threshold Technology later developed the first real ASR product called the VIP-100 System [1] for some limited application and Carnegie Mellon University under ARPA developed Harpy system, which was able to recognize speech using a vocabulary of 1,011 words with reasonable accuracy. The Harpy system was the first to take advantage of a finite-state network to reduce computation and efficiently determine the closest matching string [16]. DRAGON system by Baker [17] was also developed during the 1970s. In the 1980s, speech recognition turned toward prediction. Speech recognition vocabulary improved from about a few hundred words to several thousand words and had the potential to recognize an unlimited number of words. The major reason for this upgradation is the new statistical method HMM. Rather than simply using templates for words and looking for sound patterns, HMM considered the probability of unknown sounds being words. The foundations of modern HMM-based continuous speech recognition technology were laid down in the 1970s by groups at Carnegie- Mellon and IBM who introduced the use of discrete density HMMs [16–18] and then later at Bell Laboratories [19–21] where continuous density HMMs were introduced. Another reason of this drastic improvement of the speech recognition technology is the application of fundamental pattern recognition technology to speech recognition based on LPC methods in the mid-1970s by Itakura [22], Rabiner et al. [23], and others. Due to the expanded vocabulary provided by HMM methodology and the computer with faster processor, in the 1990s speech recognition software becomes commercially available.

During the 1980s, ANN technology was also introduced in the domain of speech recognition. The brains impressive superiority at a wide range of cognitive skills like speech recognition has motivated the researchers to explore the possibilities of ANN models in the field of speech recognition in the 1980s [24], with a hope that human neural network like models may ultimately lead to human-like performance. Early attempts at using ANNs for speech recognition centered on simple tasks like recognizing a few phonemes or a few words or isolated digits, with good success [25–27], using pattern mapping by MLP. But at the later half of 1990, suddenly ANN-based speech research got terminated [24] after the statistical framework HMM come into focus, which supports both acoustic and temporal modeling of speech. However, it should be mentioned that the current best systems are far from equaling human-like performance and many important research issues are still to be explored. Therefore, the value of ANN-based research is still at large, and nowadays, it is considered as the hot field in the domain of speech recognition.

5.3 Review of Speech Recognition During the Last Two Decades

All over the world, including India, several researches have been going on, primarily after 1990, to improve the speech recognition success rate and also to derive new methodologies. In this section, a review of a few significant works reported during the last two decades are presented. A large portion of these works are based on neuro-computing principles.

1. Rigoll reported a work in 1995 that focuses on a new ANN paradigm, and its application to recognition of speech patterns is presented. The novel ANN paradigm is a multilayer version of the well-known LVQ algorithm from Kohonen. The work gives more than five important modification to the old one [28]. Using this new technique, the work got superior result in speech recognition than the earlier LVQ.
2. In 1996, Choi et al. [29] reported a work on speech recognition using an enhanced fuzzy vector quantization (FVQ) based on a codeword-dependent distribution normalization and codeword weighting by fuzzy objective function, which presents a new variant of parameter estimation methods for discrete HMM in speech recognition. This method makes use of a codeword-dependent distribution normalization (CDDN) and a distance weighting by fuzzy contribution in dealing with the problems of robust state modeling in a FVQ-based modeling. The proposed method is compared with the existing techniques, and superior result is obtained.
3. Another work on speech recognition using the modulation spectrogram reported by Kingsbury et al. [30] in 1998. A modulation spectrogram is developed to

emphasize on the temporal structure that appears as low-frequency (below 16 Hz) amplitude modulations in subband channels following critical band frequency analysis. Visual displays of speech produced with the modulation spectrogram are relatively stable in the presence of high levels of background noise and reverberation. The modulation spectrogram is used in the front end of ASR that provides a significant improvement in performance on highly reverberant speech. When the modulation spectrogram is used in combination with log-RASTA-PLP (log RelAtive SpecTrAl Perceptual Linear Predictive analysis), performance over a range of noisy and reverberant conditions is significantly improved, which suggest that the use of multiple representations is another promising method for improving the robustness of ASR systems.

4. Deng et al. [31] reported a work on large-vocabulary speech recognition in 2000, where they have developed a new model learning paradigm that comprises a noise-insertion process followed by noise reduction and a noise adaptive training algorithm that integrates noise reduction into probabilistic multistyle system training. Further, they have developed a new algorithm (SPLICE) for noise reduction that makes no assumptions about noise stationarity and evaluated on a large-vocabulary speech recognition task. Significant and consistent error rate reduction is observed in the work.
5. Wessel et al. [32] in 2001 reported several confidence measures for large-vocabulary continuous speech recognition. They have estimated the confidence of a hypothesized word directly as its posterior probability, given all acoustic observations of the utterance. These probabilities are computed on word graphs using a forward–backward algorithm. They have also studied the estimation of posterior probabilities on best lists instead of word graphs and compare both algorithms in detail and compared the posterior probabilities with two alternative confidence measures, i.e., the acoustic stability and the hypothesis density. Experimental results are reported on five different corpora. It has been reported that the posterior probabilities computed on word graphs out perform all other confidence measures. The relative reduction in confidence error rate ranges between 19 and 35 % compared to the baseline confidence error rate.
6. Huo and Lee [33] in 2001 developed a robust speech recognition system based on adaptive classification and decision strategies. They have examined the key research issues in adaptively modifying the conventional plug-in MAP decision rules in order to improve the robustness of the classification and decision strategies used in ASR systems.
7. In 2002, a work is reported by Cowling and Sitte [34], where they have used speech recognition techniques in non-speech sound recognition where LVQ and ANN are used. They also attempted to find out the best technique among LVQ and ANN for non-speech sound recognition.
8. Another work on speech recognition with missing data using RNN has been reported by Parveen and Green in 2002, where they have proposed a missing data approach to improve the robustness of ASR to added noise. An initial process identifies spectral temporal regions, which are dominated by the speech source [35]. In this work, they have reported the development of a connectionist

approach to the problem of adapting speech recognition to the missing data case, using RNNs. In contrast, to methods based on HMMs, RNNs allow them to make use of long-term time constraints and to make the problems of classification with incomplete data and imputing missing values interact.

9. Li and Stern [36] reported a work in 2003 on feature generation based on maximum classification probability for improved speech recognition. Feature generation process is based on linear transformation of the original log-spectral representation. In this work, they also presented a comparative study of this new technique with three popular techniques, MFCC, principal component analysis (PCA), and linear discriminant analysis (LDA). Experimental results show that the new method decreases the relative word error rate compared to the best implementation of LDA and MFCC features.
10. In 2003, Povey [37] in his doctoral thesis describes about the use of discriminative criteria such as maximum mutual information (MMI) and a new minimum phone error (MPE) criteria to train HMM parameters. He has investigated the practical issues relating to the use of MMI for large-vocabulary speech recognition. The implementation gives good improvements in word error rate (WER) in case Switchboard, Broadcast News, and American Business News. The new discriminative criteria called MPE is a smoothed measure of phone transcription error. In the thesis, methods are also described for the optimization of the MPE objective function.
11. Another work reported by Ahmad et al. in 2004 where they have used RNN with BPTT for speech recognition. They have used a fully connected hidden layer between the input and state nodes and the output. The work also investigated difference between LPCC and MFCC in feature extraction process [38].
12. Another work has developed by Ala-Keturi [39] in 2004 where HMMs and ANNs are used together in speech recognition. The HMM is used with ANN because the combined system can enhance the discrimination ability.
13. Selecting appropriate feature is especially important to achieve high-speech recognition accuracy. Although the mel-cepstrum is a popular and effective feature for speech recognition, it is still unclear that the filter-bank in the mel-cepstrum is always optimal regardless of speech recognition environments or the characteristics of specific speech data. Hence, another work [40] on speech recognition describes the data-driven filter-bank optimization for a new feature extraction where they use the Kullback–Leibler (KL) distance as the measure in the filter-bank design and error rate is reduced. The work was developed by Suh and Kim in 2004.
14. Halavati et al. reported a novel approach to speech recognition using fuzzy modeling in 2004. In this work, speech spectrograms are converted into a linguistic description based on arbitrary colors and lengths. Phonemes are also described using these fuzzy measures, and recognition is done by normal fuzzy reasoning. A genetic algorithm optimizes phoneme definitions so that to classify samples into correct phonemes [41].
15. A work in 2005 reported by Yousefian and Analoui presents an approach to check the applicability of a special model of radial basis probabilistic neural

networks (RBPNN) as a classifier for speech recognition. This type of network is a combination of radial basis function (RBF) and PNN that applies characteristics of both networks and finally uses a competitive function for computing final result. The proposed network has been tested on Persian one digit numbers data set and produced significantly lower recognition error rate in comparison with other common pattern classifiers [42].

16. In 2005, a work reported by Jorgensen and Binsted [43] has developed a system to control a modified Web browser interface using subvocal electromyogram (EMG) signal classification. Here, recorded surface signals from the larynx and sublingual areas below the jaw are filtered and transformed into features using a complex dual quad tree wavelet transform. The subvocal signals are classified and used to initiate Web browser queries through a matrix-based alphabet coding scheme.
17. Scheme et al. in 2007 have reported a work on phoneme-based speech recognition, where a secondary information source is exploited using surface myoelectric signals (MES) collected from facial articulatory muscles during speech. Words are classified at the phoneme level using a HMM classifier. A fused acoustic–myoelectric multiexpert system, without knowledge of signal to noise ratio (SNR), improved on acoustic classification results at all noise levels [44].
18. Maheswari et al. developed a speech recognition system based on phonemes using ANNs in 2009, which focuses on implementing a two-module speaker-independent speech recognition system for all-British English speech. The first module performs phoneme recognition using two-level ANNs. The second module executes word recognition from the string of phonemes employing HMM [45].
19. In 2010, an work on isolated word speech recognition has reported by Savage et al. [46] where they have combined speech recognition techniques based on VQ together with ANN. VQ-based speech recognition is suitable for isolated words speech recognition for small vocabularies (100–200). It is popular due to its easy implementation and its fast calculation. To improve the performance ANN block is combined with the VQ block in this work.
20. In a work, Hawickhorst and Zahorian [47] compared three ANN architectures feedforward perceptrons trained with backpropagation, RBF networks and LVQ for ASR for speech based applications.

5.4 Research of Speech Recognition in Indian Languages

The current speech researchers are focused on using technology to overcome the challenges in natural language processing, so that next-generation speech recognition system can provide easy and natural modes of interaction for a range of applications. Specifically, it has become the primary concern for the scientists and engineers to build systems that can be consumed by the common public and to facilitate natural language transactions between human and machine. The language-specific speech recognition is difficult mainly because the system requires knowledge of word meaning, communication context, and the commonsense. This variability includes the effect of the phonetic, phonology, syntax, semantic, and communication mode of the speech signal. While having the different meaning and usage patterns, words can have the same phonetic realization. If the words were always produced in the same way, speech recognition would be relatively easy. However, for various reasons, words are almost always pronounced differently due to which it is still a challenge to build a recognizer that is robust enough in case of any speaker, any language and any speaking environment.

India is the country where vast cultural and linguistic variations are observed. Therefore, in such a multilingual environment, there is a huge possibility of implementing speech technology. The constitution of India has recognized 17 regional languages (Assamese, Bengali, Bodo, Dogri, Gujarati, Kannada, Kashmiri, Konkani, Maithili, Malayalam, Manipuri, Marathi, Nepali, Oriya, Punjabi, Sanskrit, Santhali, Sindhi, Tamil, Telugu, Urdu) along with Hindi which is the national language of India. However, till date the amount of work done in speech recognition in Indian languages has not reached the domain of rural and computer illiterate people of India [48]. Few attempts has been made by HP Labs India, IBM research laboratories, and some other research groups. Yet there is lots of scopes and possibilities to be explored to develop speech recognition system using Indian languages.

After the commercial availability of speech recognition system, the HMM technology has dominated the speech research. HMMs lie at the heart of virtually all modern speech recognition systems. The basic HMM framework has not changed significantly in the last decades, but various modeling techniques have been developed within this framework that has make the HMM technology considerably sophisticated [14, 15, 22]. At the same time from the last one or two decades, ANN technology has also been used by various researchers. The current state has considered that HMM has given the best it could, but in order to improve the accuracy of speech recognition technology under language, speaker, and environmental variations, other technology is required. In the Indian language scenario, the speech recognition work can be reviewed in two parts: work done using statistical framework like HMM, Gaussian mixture models (GMM) and a very few work done using ANN technology. Further, a few hybrid technology-based work is also found in the literature. The following sections describes the speech recognition work developed in Indian languages over the last decade.

5.4.1 Statistical Approach

The basic statistical method used in speech recognition purpose is the HMM methodology. HMMs are parametric model that can be used to model any time series but particularly suitable to model speech event. In HMM-based speech recognition, it is assumed that the sequence of observed speech vectors corresponding to each word is generated by a Markov model [49]. The forward–backward re-estimation theory

called Baum–Welch re-estimation used in HMM-based speech recognition modifies the parameter in every iteration and the probability of training data increases until a local maxima reach. The success of HMM technology lies on its capability to estimate a extended set of unknown utterance from a known set of utterance given as training set [2, 50, 51]. The availability of well-structured software like hidden Markov model tool kit (HTK) [50] and CMUs Sphinx [51], which can successfully implement HMM technology makes it easier for further research and development to incorporate new concepts and algorithms in speech recognition.
A few relevant work done in Indian languages using statistical framework like HMM are discussed below.

1. In the journal Sadhana in 1998, a work was reported by Samudravijaya et al. [52], where they have presented a description of a speech recognition system for Hindi. The system follows a hierarchic approach to speech recognition and integrates multiple knowledge sources within statistical pattern recognition paradigms at various stages of signal decoding. Rather than making hard decisions at the level of each processing unit, relative confidence scores of individual units are propagated to higher levels. A semi-Markov model processes the frame level outputs of a broad acoustic maximum likelihood classifier to yield a sequence of segments with broad acoustic labels. The phonemic identities of selected classes of segments are decoded by class-dependent ANNs, which are trained with class-specific feature vectors as input. Lexical access is achieved by string matching using a dynamic programming technique. A novel language processor disambiguates between multiple choices given by the acoustic recognizer to recognize the spoken sentence. The database used for this work consisted of a sentences having 200 words, which are most commonly used in railway reservation enquiry task.
2. Another work by Rajput et al. [53] from IBM India Research Laboratory has been reported in 2000, where they have attempted to build decision trees for modeling phonetic context dependency in Hindi by modify a decision tree built to model context dependency in American English. In a continuous speech recognition system, it is important to model the context-dependent variations in the pronunciations of phones. Linguistic–phonetic knowledge of Hindi is used to modify the English phone set. Since the Hindi phone set being used is derived from the English phone set, the adaptation of the English tree to Hindi follows naturally. The method may be applicable for adapting between any two languages.
3. In 2008, Kumar et al. of IBM India Research Laboratory developed another HMM-based large-vocabulary continuous speech recognition system for Hindi language. In this work [54], they have presented two new techniques that have been used to build the system. Initially, a technique for fast bootstrapping of initial phone models of a new language is given. The training data for the new language are aligned using an existing speech recognition engine for another language. This aligned data are used to obtain the initial acoustic models for the phones of the new language. Following this approach, less training data are required. They have also presented a technique for generating baseforms,

i.e., phonetic spellings for phonetic languages such as Hindi. As is inherent in phonetic languages, rules generally capture the mapping of spelling to phonemes very well. However, deep linguistic knowledge is required to write all possible rules, and there are some ambiguities in the language that are difficult to capture with rules. On the other hand, pure statistical techniques for baseform generation require large amounts of training data, which is not readily available. But here they have proposed a hybrid approach that combines rule-based and statistical approaches in a two-step fashion.

4. For Hindi language, Gaurav et al. have reported another work recently in 2012 [55]. A continuous speech recognition system in Hindi is tailored to aid teaching geometry in primary schools. They have used the MFCC as speech feature parameters and HMM to model these acoustic features. The Julius recognizer which is language independent was used for decoding.
5. Kumar et al. [56] in 2012 has designed a feature extraction modules ensemble of MFCC, LPCC, PLP, etc., to improve Hindi speech recognition system. The outputs of the ensemble feature extraction modules have been combined using voting technique ROVER.
6. Bhuvanagirir and Kopparapu [57] have reported another work on mixed language speech recognition Hindi and English combination in 2012.
7. In 2008, Thangarajan et al. [58] have reported a work in continuous speech recognition for Tamil language. They have build a HMM-based continuous speech recognizer based on word and triphone acoustic models. In this experiment, a word-based context-independent (CI) acoustic model for 371 unique words and a triphone-based context-dependent (CD) acoustic model for 1,700 unique words have been built for Tamil language. In addition to the acoustic models, a pronunciation dictionary with 44 base phones and trigram-based statistical language model have also been built as integral components of the linguist. These recognizers give satisfactory word accuracy for trained and test sentences read by trained and new speakers.
8. In 2009, Kalyani and Sunithato worked toward the development of a dictation system like Dragon for Indian languages. In their paper [59], they have focused on the importance of creating speech database at syllable units and identifying minimum text to be considered while training any speech recognition system. They have also provided the statistical details of syllables in Telugu and its use in minimizing the search space during recognition of speech. The minimum words that cover maximum syllables are identified, which can be used for preparing a small text for collecting speech sample while training the dictation system.
9. Another work in Telugu language is reported by Usha Rani and Girija in 2012. To improve the speech recognition accuracy on Telugu language, they have explored means to reduce the number of the confusion pairs by modifying the dictionary, which is used in the decoder of the speech recognition system. In their paper [60], they have described the different types of errors obtained from the decoder of the speech recognition system.
10. Das et al. [61] reported a work in Bengali language, where they have described the design of a speech corpus for continuous speech recognition. They have

developed speech corpora in phone and triphone labeled between two age groups—20 to 40 and 60 to 80. HMM is used to align the speech data statistically and observed good performance in phoneme recognition and continuous word recognition done using HTK and SPHINX.

11. In Punjabi language, Dua et al. [62] have reported a work in 2012, where they have attempted to develop a isolated word recognition system using HTK.
12. In 2013, another work has been reported by Mehta et al. [63] where a comparative study of MFCC and LPC for Marathi isolated word recognition system is described.
13. Udhyakumar et al. [64] reported a work in 2004 for multilingual speech recognition to be used for information retrieval in Indian context. This paper analyzes various issues in building a HMM-based multilingual speech recognizer for Indian languages. The system is designed for Hindi and Tamil languages and adapted to incorporate Indian accented English. Language-specific characteristics in speech recognition framework are highlighted. The recognizer is embedded in information retrieval applications, and hence, several issues like handling spontaneous telephony speech in real-time, integrated language identification for interactive response and automatic grapheme to phoneme conversion to handle out of vocabulary words are addressed in this paper.
14. Some issues about the development of speech databases of Tamil, Telugu, and Marathi for large-vocabulary speech recognition system is reported in a work [65] by Anumanchipalli et al. in 2005. They have collected speech data from about 560 speakers in these three languages. They have also presented the preliminary speech recognition results using the acoustic models created on these databases using Sphinx 2 speech tool kit, which shows satisfactory improvement in accuracy.
15. During 2009, Murthy et al. [66, 67] described in a novel approach to build syllable-based continuous speech recognizers for Indian languages, where a syllable-based lexicon and language model are used to extract the word output from the HMM-based recognizer. The importance of syllables as the basic subword unit for recognition has been a topic for research. They have shown that a syllabified lexicon helps hypothesize the possible set of words, where the sentence is constructed with the help of N-gram-based statistical language models. The database used for these works is the Doordarshan database, which is made of news bulletins of approximately 20 min duration of both male and female speakers.
16. Bhaskar et al. [68] reported another work on multilingual speech recognition in 2012. They have used a different approach to multilingual speech recognition, in which the phone sets are entirely distinct, but the model has parameters not tied to specific states that are shared across languages. They have tried to build a speech recognition system using HTK for Telugu language. The system is trained for continuous Telugu speech recorded from native speakers.
17. In 2013, a work has been reported by Mohan et al. [69] where a spoken dialog system is designed to use in agricultural commodities task domain using real-world speech data collected from two linguistically similar languages of

India, Hindi and Marathi. They have trained a subspace Gaussian mixture model (SGMM) under a multilingual scenario [70, 71]. To remove acoustic, channel and environmental mismatch between data sets from multiple languages, they have used a cross-corpus acoustic normalization procedure which is a simple variant of speaker adaptive training (SAT) described by Mohan et al. [72] in 2012. The resulting multilingual system provides the best speech recognition performance of 77.77 % for both languages .

It is observed from the above literature that the speech recognition technology in India is dominated by the statistical approaches like HMM and GMM and the focus is on improving the success rate of continuous speech recognition in Indian languages. A few researchers have also focused on the linguistic and phonetic diversity of the sounds of Indian languages so that the robustness of the ASR technology can be improved.

5.4.2 ANN-Based Approach

All the above-mentioned works available in open literature are based on HMM technology. All earlier theories of spoken word recognition [73–77] agree to the fact that the spoken word recognition is a complex multileveled pattern recognition work performed by neural networks of human brain and the related speech perception process can be modeled as a pattern recognition network. Different levels of the language like lexical, semantic, phonemes can be used as the unit of distribution in the model. All the theories proposed that bottom-up and top-down processes between feature, phoneme, and word level combine to recognize a presented word. In such a situation, ANN models have the greatest potential, where hypothesis can be performed in a parallel and higher computational approach. ANN models are composed of many nonlinear computational elements operating in parallel and arranged in the pattern of biological neural network. The problem of speech recognition inevitably requires handling of temporal variation and ANN architectures like RNN, TDNN etc. have proven to be handy in such situations. However, ANN-based speech recognition research is still growing. A few works based on ANN technology are listed below.

1. Sarkar and Yegnanarayana have used fuzzy-rough neural networks for vowel classification in a work reported in 1998. This paper [78] has proposed a fuzzy-rough set-based network that exploits fuzzy-rough membership functions while designing radial basis function (RBF) neural networks for classification.
2. In 2001, Gangashetty and Yegnanarayana have described ANN models for recognition of consonant–vowel (CV) utterances in [79]. In this paper, an approach based on ANN models for recognition of utterances of syllable like units in Indian languages is described. The distribution capturing ability of an autoassociative neural network (AANN) model is exploited to perform nonlinear principal component analysis (PCA) for compressing the size of the feature vector.

A constraint satisfaction model is proposed in this paper to incorporate the acoustic–phonetic knowledge and to combine the outputs of subnets to arrive at the overall decision on the class of an input utterance.

3. Khan et al. [80] described an ANN-based preprocessor for recognition of syllables in 2004. In this work, syllables in a language are grouped into equivalent classes based on their consonant and vowel structure. ANN models are used to preclassify the syllables into the equivalent class to which they belong. Recognition of the syllables among the smaller number of cohorts within a class is done by means of HMMs. The preprocessing stage limits the confusable set, to the cohorts within a class and reduces the search space. This hybrid approach helps to improve the recognition rate over that of a plain HMM-based recognizer.
4. In 2005, Gangashetty et al. have described an approach for multilingual speech recognition by spotting consonant–vowel (CV) units. The distribution capturing capability of AANN is used for detection of vowel onset points in continuous speech. Support Vector Machine (SVM) classifier is used as classifier, and broadcast news corpus of three Indian languages Tamil, Telugu, and Marathi is used [81].
5. Paul et al. [82] in 2009 reported a work on Bangla speech recognition using LPC cepstrum features. The SOM structure of ANN makes each variable length LPC trajectory of an isolated word into a fixed-length LPC trajectory and thereby making the fixed-length feature vector to be fed into the recognizer. The structures of the ANN is designed with MLP and tested with 3, 4, 5 hidden layers using the tan-sigmoid transfer functions.
6. In 2012, Sunil and Lajish [83] in a work reported a model for vowel phoneme recognition based on average energy information in the zero crossing intervals and its distribution using multilayer feedforward ANN. They have observed that the distribution patterns of average energy in the zero crossing intervals are similar for repeated utterances of the same vowels and vary from vowel to vowel and this parameter is used as feature to classify five Malayalam vowels in the recognition system. From this study, they have shown that the average energy information in the zero crossing intervals and its distributions can be effectively utilized for vowel phone classification and recognition.
7. Pravin and Jethva [84] recently in 2013 reported a work on Gujrati speech recognition. MFCC of a few selected spoken words is used as feature to train a MLP-based word recognition system.
8. Chitturi et al. [85] reported a work in 2005, where they have proposed an ANN-based approach to model the lexicon of the foreign language with a limited amount training data. The training data for this work consisted of the foreign language with the phone set of three native languages, 49 phones in Telugu, 35 in Tamil, 48 in Marathi, and 40 in US English. The MLP with backpropagation learning algorithm learns how the phones of the foreign language vary with different instances of context. The trained network is capable of deciphering the pronunciation of a foreign word given its native phonetic composition. The performance of the technique has been tested by recognizing Indian accented English.

9. In a work by Thasleema and Narayanan [86] in 2012, explores the possibility of multiresolution analysis for consonant classification in noisy environments. They have used wavelet transform (WT) to model and recognize the utterances of consonant–vowel (CV) speech units in noisy environments. A hybrid feature extraction module, namely Normalized Wavelet Hybrid Feature (NWHF) using the combination of Classical Wavelet Decomposition (CWD) and Wavelet Packet Decomposition (WPD) along with z-score normalization technique, is designed in this work. CV speech unit recognition tasks performed for both noisy and clean speech units using ANN and k nearest neighborhood.
10. In 2010, Sukumar et al. [87] have reported a work on recognition of isolated question words of Malayalam language from speech queries using ANN and discrete wavelet transform (DWT)-based speech feature.
11. Sarma et al. have reported works on recognition of numerals of Assamese language in [88, 89] using ANN in 2009. In [88], the ANN models are designed using a combination of SOM and MLP constituting a LVQ block trained in a cooperative environment to handle male and female speech samples of numerals. This work provides a comparative evaluation of several such combinations while subjected to handle speech samples with gender-based differences captured by a microphone in four different conditions, viz. noiseless, noise mixed, stressed, and stress-free. In [89], the effectiveness of using an adaptive LMS filter and LPC cepstrum to recognize isolated numerals using ANN-based cooperative architectures is discussed. The entire system has two distinct parts for dealing with two classes of input classified into male and female clusters. The first block is formed by a MLP which acts like a class mapper network. It categories the inputs into two gender-based clusters.
12. In 2010, Bhattacharjee [90] presented a technique for the recognition of isolated keywords from spoken search queries. A database of 300 spoken search queries from Assamese language has been created. In this work, MFCC has been used as the feature vector and MLP to identify the phoneme boundaries as well as for recognition of the phonemes. Viterbi search technique has been used to identify the keywords from the sequence of phonemes generated by the phoneme recognizer.
13. In a work by Dutta and Sarma [91] in 2012, they describe a speech recognition model using RNN where LPC and MFCC are used for feature extraction in two separate decision block and decision is taken from the combined architecture. The multiple feature extraction block-based model provides 10 % gain in the recognition rate in comparison with the case when individual feature extractor is used.
14. In 2013, Bhattacharjee in [92] provided a comparative study of LPCC and MFCC features for the recognition of phones of Assamese language. Two popular feature extraction techniques, LPCC and MFCC, have been investigated, and their performances have been evaluated for the recognition using a MLP-based baseline phoneme recognizer in quiet environmental condition as well as at different level of noise. It has been reported that the performance of LPCC-based system degrades more rapidly compare to MFCC-based system under environ-

mental noise condition, whereas under speaker variability conditions, LPCC shows relative robustness compare to MFCC though the performance of both the systems degrades considerably.
15. Sarma and Sarma [93] in 2013 reported a work where a hybrid ANN model is used to recognize initial phones from CVC-type Assamese words. A SOM-based algorithm is developed to segment the initial phonemes from its word counterpart. Using a combination of three types of ANN structures, namely RNN, SOM, and PNN, the proposed algorithm proves its superiority over the DWT-based phoneme segmentation.

It is observed that ANN-based technology is also adopted by a few Indian researcher. But the use of ANN is limited to isolated word level or phone level.

5.5 Conclusion

In this chapter, we have described the available literatures covering the advances of speech recognition technology observed in the last two decades. During the initial stages, speech recognition technology adopted certain ANN-based approaches including feedforward topologies and LVQ methods. Later, it shifted toward HMM and other similar statistical tools. A few works have also used hybrid blocks containing both learning and HMM-based methods. The objective has been to optimize design requirements and improve performance. It can also be concluded from the literature that the speech recognition technology for Indian languages has not yet covered all the official languages. A few works are done in Hindi language by IBM research laboratory, and a few other research groups have appreciable quality. A few other works have covered Marathi, Tamil, Telugu, Punjabi, Assamese, and Bengali languages, which are widely spoken throughout the country. A few works are reported on multilingual speech recognition. However, ASR technologies are yet to be reported in some other mainstream language like Urdu, Sanskrit, Kashmiri, Sindhi, Konkani, Manipuri, Kannada, and Nepali. The HMM-based works have already supported the use of continuous speech. In contrast, ANN-based works are still centered around isolated words. But the scenario looks bright and many new success stories shall be reported in near future which shall take ASR technology in Indian languages to new heights.

References

1. Juang BH, Rabiner LR (2004) Automatic speech recognition: a brief history of the technology development. Available via http://www.ece.ucsb.edu/Faculty/Rabiner/ece259/Reprints/354_LALI-ASRHistory-final-10-8.pdf
2. Gales M, Young S (2007) The application of hidden Markov models in speech recognition. Found Trends Sig Process 1(3):195–304

3. Levinson SE, Rabiner LR, Sondhi MM (1983) An introduction to the application of the theory of probabilistic functions of a Markov process to automatic speech recognition. Bell Sys Tech J 62(4):1035–1074
4. Ferguson JD (1980) Hidden Markov analysis: an introduction, in hidden Markov models for speech. Princeton, Institute for Defense Analyses
5. Rabiner LR, Juang BH (2004) Statistical methods for the recognition and understanding of speech. Encyclopedia of language and linguistics. Available via http://www.idi.ntnu.no/~gamback/teaching/TDT4275/literature/rabiner_juang04.pdf
6. Davis KH, Biddulph R, Balashek S (1952) Automatic recognition of spoken digits. J Acous Soc Am 24(6):637–642
7. Suzuki J, Nakata K (1961) Recognition of japanese vowels preliminary to the recognition of speech. J Radio Res Lab 37(8):193–212
8. Sakai J, Doshita S (1962) The phonetic typewriter. In: Information processing. Proceedings, IFIP congress, Munich
9. Nagata K, Kato Y, Chiba S (1963) Spoken digit recognizer for Japanese language. NEC research and development, 6
10. Fry DB, Denes P (1959) The design and operation of the mechanical speech recognizer at university college london. J British Inst Radio Eng 19(4):211–229
11. Martin TB, Nelson AL, Zadell HJ (1964) Speech recognition by feature abstraction techniques. Technical report AL-TDR-64-176, Air Force Avionics Laboratory
12. Vintsyuk TK (1968) Speech discrimination by dynamic programming. Kibernetika 4(2):81–88
13. Viterbi AJ (1967) Error bounds for convolutional codes and an asymptotically optimal decoding algorithm. IEEE Trans Inf Theor 13:260–269
14. Atal BS, Hanauer SL (1971) Speech analysis and synthesis by linear prediction of the speech wave. J Acoust Soc Am 50(2):637–655
15. Itakura F (1970) A statistical method for estimation of speech spectral density and formant frequencies. Electron Commun Jpn 53(A):36–43
16. Lowerre BT (1976) The HARPY speech recognition system. Doctoral thesis, Carnegie-Mellon University, Department of Computer Science
17. Baker JK (1975) The dragon system: an overview. IEEE Trans Acoust Speech Sig Process 23(1):24–29
18. Jelinek F (1976) Continuous speech recognition by statistical methods. Proc IEEE 64(4):532–556
19. Juang BH (1984) On the hidden Markov model and dynamic time warping for speech recognition: a unified view. AT&T Bell Lab Tech J 63(7):1213–1243
20. Juang BH (1985) Maximum-likelihood estimation for mixture multivariate stochastic observations of Markov chains. AT&T Bell Lab Tech J 64(6):1235–1249
21. Levinson SE, Rabiner LR, Sondhi MM (1983) An introduction to the application of the theory of probabilistic functions of a Markov process to automatic speech recognition. Bell Syst Tech J 62(4):1035–1074
22. Itakura F (1975) Minimum prediction residual principle applied to speech recognition. IEEE Trans Acoust Speech Sig Process 23:57–72
23. Rabiner LR, Levinson SE, Rosenberg AE, Wilpon JG (1979) Speaker independent recognition of isolated words using clustering techniques. IEEE Trans Acoust Speech Sig Process 27:336–349
24. Hu YH, Hwang JN (2002) Handbook of neural network signal processing. The electrical engineering and applied signal processing series. CRC Press, USA
25. Lippmann RP (1990) Review of neural networks for speech recognition. Readings in speech recognition. Morgan Kaufmann Publishers, Burlington, pp 374–392
26. Evermann G, Chan HY, Gales MJF, Hain T, Liu X, Mrva D, Wang L, Woodland P (2004) Development of the 2003 CU-HTK conversational telephone speech transcription system. In: Proceedings of ICASSP, Montreal, Canada

27. Matsoukas S, Gauvain JL, Adda A, Colthurst T, Kao CI, Kimball O, Lamel L, Lefevre F, Ma JZ, Makhoul J, Nguyen L, Prasad R, Schwartz R, Schwenk H, Xiang B (2006) Advances in transcription of broadcast news and conversational telephone speech within the combined ears bbn/limsi system. IEEE Trans Audio Speech Lang Process 14(5):1541–1556
28. Rigoll G (1995) Speech recognition experiments with a new multilayer LVQ network (MLVQ). In: Proceedings of eurospeech, ISCA
29. Choi HJ, Ohand YH, Dong YK (1996) Speech recognition using an enhanced FVQ based on a codeword dependent distribution normalization and codeword weighting by fuzzy objective function. In: Proceedings of the international conference on spoken language processing
30. Kingsbury BED, Morgan N, Greenberg S (1998) Robust speech recognition using the modulation spectrogram. Speech Commun 25:117–132
31. Deng L, Acero A, Plumpe M, Huang X (2000) Large-vocabulary speech recognition under adverse acoustic environments. In: Proceedings of interspeech, ISCA, pp 806–809
32. Wessel F, Schluter R, Macherey K, Ney H (2001) Confidence measures for large vocabulary continuous speech. IEEE Trans Speech Audio Process 9(3):288–298
33. Huo Q, Lee CH (2001) Robust speech recognition based on adaptive classification and decision strategies. Speech Commun 34:175–194
34. Cowling M, Sitte R (2002) Analysis of speech recognition techniques for use in a non-speech sound recognition system. In: Proceedings of 6th international symposium on digital signal processing for communication systems
35. Parveen S, Green PD (2002) Speech recognition with missing data using recurrent neural nets. In: Proceedings of the 14th conference. Advances in neural information processing systems, vol 2, pp 1189–1194
36. Li X, Stern RM (2003) Feature generation based on maximum classification probability for improved speech recognition. Interspeech, ISCA
37. Povey D (2003) Discriminative training for large vocabulary speech recognition. Doctoral thesis, University of Cambridge, Cambridge
38. Ahmad AM, Ismail S, Samaonl DF (2004) Recurrent neural network with backpropagation through time for speech recognition. In: Proceedings of international symposium on communications and information technologies, Sapporo, Japan
39. Ala-Keturi V (2004) Speech recognition based on artificial neural networks. Helsinki University of Technology, Available via http://www.cis.hut.fi/Opinnot/T-61.6040/pellom-2004/project-reports/project_07.pdf
40. Suh Y and Kim H (2004) Data-driven filter-bank-based feature extraction for speech recognition. In: Proceedings of the 9th conference speech and computer, St. Petersburg, Russia
41. Halavati R, Shouraki S B, Eshraghi M, Alemzadeh M (2004) A novel fuzzy approach to speech recognition. In: Proceedings of 4th international conference on hybrid intelligent systems
42. Yousefian N, Analoui M (2005) Using radial basis probabilistic neural network for speech recognition. Available via http://confbank.um.ac.ir/modules/conf_display/conferences/ikt07/pdf/F3_3.pdf
43. Jorgensen C, Binsted K (2005) Web browser control using EMG based sub vocal speech recognition. In: Proceedings of the 38th annual Hawaii international conference on system Sciences, p 294c
44. Scheme EJ, Hudgins B, Parker PA (2007) Myoelectric signal classification for phoneme-based speech recognition. IEEE Trans Biomed Eng 54(4):694–699
45. Maheswari NU, Kabilan AP, Venkatesh R (2009) Speech recognition system based on phonemes using neural networks. Int J Comput Sci Netw Secur 9(7):148–153
46. Savage J, Rivera C, Aguilar V (2010) Isolated word speech recognition using vector quantization techniques and artificial neural networks
47. Hawickhorst BA, Zahorian SA, A comparison of three neural network architectures for automatic speech recognition. Department of Electrical and Computer Engineering Old Dominion University, Norfolk. Available via http://www.ece.odu.edu/szahoria/pdf/
48. Languages with official status in India. The Constitution of India, eighth schedule, articles 344(1) and 351: 330. Available via http://lawmin.nic.in/coi/coiason29july08.pdf

49. Rabiner L, Juang BH (1986) An introduction to hidden Markov models. IEEE ASSP Mag 3(1):4–16
50. Young S, Kershaw D, Odell J, Ollason D, Valtchev V, Woodland P (2000) The HTK book. Available via http://htk.eng.cam.ac.uk/
51. Lee KF, Hon HW, Reddy R (1990) An overview of the SPHINX speech recognition system. IEEE Trans Acoust Speech Sig Process 38(1):35–45
52. Samudravijaya K, Ahuja R, Bondale N, Jose T, Krishnan S, Poddar P, Rao PVS, Raveendran R (1998) A feature-based hierarchical speech recognition system for hindi. Sadhana 23(4):313–340
53. Rajput N, Subramaniam LV, Verma A (2000) Adapting phonetic decision trees between languages for continuous speech recognition. In: Proceedings of IEEE international conference on spoken language processing, Beijing, China
54. Kumar M, Rajput N, Verma A (2004) A large-vocabulary continuous speech recognition system for Hindi. IBM J Res Dev 48(5/6):703–715
55. Gaurav DS, Deiv G, Sharma K, Bhattacharya M (2012) Development of application specific continuous speech recognition system in hindi. J Sig Inf Process 3:394–401
56. Kumar M, Aggarwal RK, Leekha G, Kumar Y (2012) Ensemble feature extraction modules for improved hindi speech recognition system. Proc Int J Comput Sci Issues 9(3):359–364
57. Bhuvanagirir K, Kopparapu SK (2012) Mixed language speech recognition without explicit identification of language. Am J Sig Process 2(5):92–97
58. Thangarajan R, Natarajan AM, Selvam M (2008) Word and triphone based approaches in continuous speech recognition for tamil language. WSEAS Trans Sig Process 4(3):76–85
59. Kalyani N, Sunitha KVN (2009) Syllable analysis to build a dictation system in telugu language. Int J Comput Sci Inf Secur 6(3):171–176
60. Usha Rani N, Girija PN (2012) Error analysis to improve the speech recognition accuracy on telugu language. Sadhana 37(6):747–761
61. Das B, Mandal S, Mitra P (2011) Bengali speech corpus for continuous automatic speech recognition system. In: Proceedings of international conference on speech database and assessments, pp 51–55
62. Dua M, Aggarwal RK, Kadyan V, Dua S (2012) Punjabi automatic speech recognition using HTK. Int J Comput Sci Issues 9(4):359–364
63. Mehta LR, Mahajan SP, Dabhade AS (2013) Comparative study of MFCC and LPC for Marathi isolated word recognition system. Int J Adv Res Electr Electron Instrum Eng 2(6):2133–2139
64. Udhyakumar N, Swaminathan R, Ramakrishnan SK (2004) Multilingual speech recognition for information retrieval in indian context. In: Proceedings of the student research workshop at HLT-NAACL, pp 1–6
65. Anumanchipalli G, Chitturi R, Joshi S, Kumar R, Singh SP, Sitaram RNV, Kishore SP (2005) Development of Indian language speech databases for large vocabulary speech recognition systems. In: Proceedings of international conference on speech and computer
66. Lakshmi A, Murthy HA (2008) A new approach to continuous speech recognition in Indian languages. In: Proceedings national conferrence communication
67. Lakshmi SG, Lakshmi A, Murthy HA, Nagarajan T (2009) Automatic transcription of continuous speech into syllable-like units for Indian languages. Sadhana 34(2):221–233
68. Bhaskar PV, Rao SRM, Gopi A (2012) HTK based Telugu speech recognition. Int J Adv Res Comput Sci Softw Eng 2(12):307–314
69. Mohan A, Rose R, Ghalehjegh SH, Umesh S (2013) Acoustic modelling for speech recognition in Indian languages in an agricultural commodities task domain. Speech Commun 56:167–180. Available via http://dx.doi.org/10.1016/j.specom.2013.07.005
70. Povey D, Burget L, Agarwal M, Akyazi P, Kai F, Ghoshal A, Glembek O, Goel N, Karafiat M, Rastrow A (2011) The subspace Gaussian mixture model: a structured model for speech recognition. Comput Speech Lang 25(2):404–439
71. Rose RC, Yin SC, Tang Y (2011) An investigation of subspace modeling for phonetic and speaker variability in automatic speech recognition. In: Proceedings of IEEE international conference on acoustics, speech, and signal processing

72. Mohan A, Ghalehjegh SH, Rose RC (2012) Dealing with acoustic mismatch for training multilingual subspace Gaussian mixture models for speech recognition. In: Proceedings of IEEE international conference on acoustics, speech and signal processing
73. Diehl RL, Lotto AJ, Holt LL (2004) Speech perception. Annu Rev Psychol 55:149–179
74. Eysenck MW (2004) Psychology-an international perspective. Psychology Press. Available via http://books.google.co.in/books/about/Psychology.html?id=l8j_z5-qZfACredir_esc=y
75. Jusczyk PW, Luce PA (2002) Speech perception and spoken word recognition: past and present. Ear Hear 23(1):2–40
76. Bergen B (2006) Linguistics 431/631: connectionist language modeling. Meeting 10: speech perception. Available via http://www2.hawaii.edu/bergen/ling631/lecs/lec10.htm
77. McClelland JL, Mirman D, Holt LL (2006) Are there interactive processes in speech perception? Trends Cogn Sci 10(8):363–369
78. Sarkar M, Yegnanarayana B (1998) Fuzzy-rough neural networks for vowel classification. IEEE international conference on systems, man, and cybernetics, p 5
79. Gangashetty SV, Yegnanarayana B (2001) Neural network models for recognition of consonant-vowel (CV) utterances. In: Proceedings of international joint conference on neural networks
80. Khan AN, Gangashetty SV, Yegnanarayana B (2004) Neural network preprocessor for recognition of syllables. In: Proceedings of international conference on intelligent sensing and information processing
81. Gangashetty SV, Sekhar CC, Yegnanarayana B (2005) Spotting multilingual consonant-vowel units of speech using neural network models. In: Proceeding of international conference on non-linear speech processing
82. Paul AK, Das D, Kamal M (2009) Bangla speech recognition system using LPC and ANN. In: Proceedings of the seventh international conference on advances in pattern recognition, pp 171–174
83. Sunil KRK, Lajish VL (2012) Vowel phoneme recognition based on average energy information in the zerocrossing intervals and its distribution using ANN. Int J Inf Sci Tech 2(6):33–42
84. Pravin P, Jethva H (2013) Neural network based Gujarati language speech recognition. Int J Comput Sci Manage Res 2(5):2623–2627
85. Chitturi R, Keri V, Anumanchipalli G, Joshi S (2005) Lexical modeling for non native speech recognition using neural networks. In: Proceedings of international conference of natural language processing
86. Thasleema TM, Narayanan NK (2012) Multi resolution analysis for consonant classification in noisy environments. Int J Image Graph Sig Process 8:15–23
87. Sukumar AR, Shah AF, Anto PB (2010) Isolated question words recognition from speech queries by using Artificial Neural Networks. In: Proceedings of international conference on computing communication and networking technologies, pp 1–4
88. Sarma MP, Sarma KK (2009) Assamese numeral speech recognition using multiple features and cooperative lvq-architectures. Int J Electr Electron Eng 5(1):27–37
89. Sarma M, Dutta K, Sarma KK (2009) Assamese numeral corpus for speech recognition using cooperative ANN architecture. World Acad Sci Eng Tech 28:581–590
90. Bhattacharjee U (2010) Search key identification in a spoken query using isolated keyword recognition. Int J Comput Appl 5(8):14–21
91. Dutta K, Sarma KK (2012) Multiple feature extraction for RNN-based assamese speech recognition for speech to text conversion application. In: Proceedings of international conference on communications, devices and intelligent systems (CODIS), pp 600–603
92. Bhattacharjee U (2013) A comparative study of LPCC And MFCC features for the recognition of assamese phonemes. Int J Eng Res Technol 2(1):1–6
93. Sarma M, Sarma KK (2013) An ANN based approach to recognize initial phonemes of spoken words of assamese language. Appl Soft Comput 13(5):2281–2291

Part II
Design Aspects

Phonemes are the smallest distinguishable unit of a speech signal. Segmentation of phoneme from its word counterpart is a fundamental and crucial part in various speech processing fields like spoken word recognition, speaker identification etc. This work presents a new phoneme segmentation algorithm based on two different supervised and unsupervised ANN structures forming a hybrid soft-computational framework which sequentially segments and classifies the phonemes of spoken words of Assamese language. The algorithm uses weight vectors, obtained by training SOM with different number of iteration, as a segment of different phonemes constituting the word whose LPC samples are used for training. Phonemes are linguistic abstraction and SOM extracts out some abstract internal representation from the speech signal. SOM trained with different iteration number provides varied internal phoneme segments. Some two class PNN based classification performed with clean Assamese vowel and consonant phonemes is used to identify these segments. The work also shows the application of the proposed segmentation technique in two different speech processing field- spoken word recognition and vowel based speaker identification.

Initial phoneme is used to activate words starting with a particular phoneme in spoken word recognition models. Investigating the initial phoneme, one can classify them into a phonetic group, and then phonemes can be classified within the group. A spoken word recognition model is described in this work, where all the words of the recognition vocabulary are initially classified into six different phoneme families of Assamese language, on the basis of its initial phoneme. Later, within the classified family consonant and vowel phonemes of the word are segmented and recognized. The SOM based segmentation method which segments the spoken words into its constituent phonemes and from the SOM segmented phonemes, the constituent phonemes are identified by some sort of recognition algorithm based on two different ANN structures namely PNN and LVQ. Formant frequency of Assamese vowel phonemes is used effectively as a priori knowledge in the vowel recognition algorithm. The SOM based segmentation algorithm shows distinct advantage in terms of word recognition success rate in comparison to the conventional speech segmentation methods like windowing or Discrete Wavelet Transform (DWT). Further a method is described using K-mean clustering (KMC), so that the spoken word recognition model can be extend to the include words having more than three phonemes.

The second application approach is the identification of Assamese speaker using the vowel sound segmented out from words spoken by a speaker. The work uses a proposed SOM and PNN based approach to segment the vowel phoneme. The segmented vowel is later used to identify the speaker of the word by matching patterns with a LVQ based speaker code book. The LVQ code book is prepared by taking LP residual features of clean vowel phonemes uttered by the male and female speakers to be identified. The proposed method formulates a framework for the design of a speaker identification model of Assamese language, which has some distinctive variation of occurrence of vowel sound in the same word spoken by speakers of different dialects.

Chapter 6
Phoneme Segmentation Technique Using Self-Organizing Map (SOM)

Abstract This chapter provides a description of the proposed SOM-based segmentation technique and explains how it can be used to segment the initial phoneme from some CVC-type Assamese word. In Sect. 6.3, the SOM and PNN have been explained in a more relevant way, as they are used as parts of the technique. The work provides a comparison of the proposed SOM-based technique with the conventional DWT-based speech segmentation technique. Therefore, some experimental works are carried out to perform the DWT-based phoneme segmentation. The theoretical details of DWT-based technique are explained in Sect. 6.4. In Sect. 6.5, algorithms associated with the proposed segmentation technique are explained along with the experimental details. Result and discussions are included in the Sect. 6.6. Section 6.7 concludes the chapter.

Keywords CVC · Segmentation · SOM · PNN · DWT

6.1 Introduction

Segmentation of data is important in many aspects of image and signal processing. The general idea is to divide the data into segments such that each segment can be treated separately to achieve specific goals such as pattern recognition, feature extraction, data compression, template matching, and noise reduction. The criterion to determine the splitting locations evidently depends on the application. Segmentation of constituent phonemes from a word is the most crucial component of any phoneme-based spoken word recognition model. The accuracy of the segmentation technique directly effects the success rate of phoneme recognition system. The most frequently used current method of speech segmentation is the constant time framing, for example, into 25 ms blocks. This method benefits from simplicity of implementation and the ease of comparing blocks of the same length. However, the varying length of phonemes is a natural phenomenon that cannot be ignored. Moreover, boundary effects provide additional distortion. Boundary effects are typically reduced by

M. Sarma and K. K. Sarma, *Phoneme-Based Speech Segmentation Using Hybrid Soft Computing Framework*, Studies in Computational Intelligence 550,
DOI: 10.1007/978-81-322-1862-3_6, © Springer India 2014

applying the Hamming window. Obviously, framing creates more boundaries than phonemes. Constant segmentation, therefore, risks losing information about the phonemes due to merging different sounds into single blocks, losing phoneme length information, and losing complexity of individual phonemes. A more satisfactory approach is an attempt to find the phoneme boundaries from the time-varying speech signal properties. A number of approaches have been previously suggested for this task [1], but these utilize features derived from acoustic knowledge of the phonemes. Such methods need to be optimized to a particular phoneme data and cannot be performed in isolation from phoneme recognition. DWT is also frequently used for such kind of phoneme segmentation [2–4]. DWT easily extracts speech parameters that take into account the properties of the human hearing system, but the recognition success rate obtained by DWT-based segmentation is not satisfactory. DWT's approach to formulating segmentation boundaries is based on a filter bank that suffers from sampling/aliasing effects and fails to generate near perfect reconstruction that too within a spectral range of around 500 Hz. It leads to distortion for which success rates suffer. DWT's limited spectral response within the recorded frequency range of the speech signals prevents formulation of segmentation boundaries, which is a straight requirement to achieve higher success rates during recognition.

This chapter presents an ANN approach to speech segmentation by extracting the weight vector obtained from SOM trained with the LPC features of digitized samples of speech to be segmented. It is observed that varying the training epochs, different phonemes can be segmented from a particular word. A PNN-based classifier trained with clean Assamese phoneme is used to classify the SOM-segmented phonemes which indirectly proves the validation of the SOM-based segmentation technique developed for the work. Assamese phonemes and words containing those phonemes are recorded from five girls and five boys, so that the classification algorithm can remove the speaker dependance limitation. Thus, in the proposed algorithm, some abstract internal form of the speech signal is extracted in terms of weight vectors obtained by training SOM. Subsequently, using the trained PNN phoneme segments are identified from those weight vectors and boundaries laid. The work shows almost 96–98 % success rate, which is well above the previously reported results. Thus, in the proposed algorithm, some abstract internal form of the speech signal is extracted in terms of weight vectors obtained by training SOM. Subsequently, using the trained PNN phoneme segments are identified from those weight vectors and boundaries laid.

6.2 Linear Prediction Coefficient (LPC) as Speech Feature

Linear prediction coding (LPC) is one of the most popularly used speech analysis technique, based on a mathematical approximation of the vocal tract. The basic idea behind linear predictive analysis is that a specific speech sample at the current time can be approximated as a linear combination of past speech samples. Through minimizing the sum of squared differences over a finite interval between the actual speech samples and linear predicted values, a unique set of parameters or predictor coefficients can be determined [5]. LPC starts with the assumption that the speech signal is produced by a buzzer at the end of a tube. The glottis is considered as the buzz, which is characterized by intensity and frequency. As mentioned earlier, the vocal tract forms the tube, which is characterized by its resonances. These are called formants. LPC analyzes the speech signal by estimating the formants, removing their effects from the speech signal, and estimating the intensity and frequency of the remaining buzz. The process of removing the formants is called inverse filtering, and the remaining signal is called the residue.

If short-term linear prediction analysis is done, then for each short-term frame a set of coefficients are obtained, which is considered as a feature vector. The feature vector must contain information that is useful to identify and differentiate speech sounds and insensitive to speaker individuality and other irrelevant factors. To capture the dynamics of the vocal tract movements, the short-term spectrum is typically computed every 10 ms using a window of 20 ms. The envelope of the spectrum can be represented indirectly in terms of the parameters of an all-pole model, using LPC. The linear prediction model of speech signal is explained in more detail in Chap. 7, with reference to the estimation of vowel formants.

6.3 Application of SOM and PNN for Phoneme Boundary Segmentation

A brief treatment of the two soft computational tools, SOM and PNN, is presented here with descriptions modified to make it as relevant as possible for the present work.

- SOM: SOM is a special form of ANN trained with unsupervised paradigm of learning. It follows a competitive method of learning which enables it to work effectively as a feature map, classifier, and, at times, as a filter. SOM has a special property of effectively creating spatially organized 'internal representations' of various features of input signals and their abstractions [6]. SOM can be considered as a data visualization technique, that is, it provides some underlying structure of the data [7]. This is the basic theoretical idea used in our vowel segmentation technique. If we train the same SOM with different epochs or iteration numbers for a particular epoch, SOM provides a weight vector consisting of the wining neuron along with the neighbors. Thus, with different epochs, different internal segments or patterns of the speech signal can be obtained. Suppose, we have a one-dimensional field of neurons and say the LPC samples of spoken word uttered by the speaker has a form

$$X_k = (x_1^k, x_2^k, \ldots, x_n^k) \tag{6.1}$$

When such an input vector is presented to the field of neurons, the algorithm will start to search the best matching weight vector W_i and thereby identify a neighborhood $\aleph_i$ around the wining neuron. While doing so, it will try to minimize the Euclidian distance $|X_k - W_i(k)|$. Adaptation of the algorithm will take place according to the relation,

$$W_i(k+1) = \begin{cases} W_i(k) + \eta_k(X_k - W_i(k)), & \mathrm{i} \in \aleph_{\mathrm{i}}^{\mathrm{k}}; \\ W_i(k), & \mathrm{i\ not\ } \aleph_i^k; \end{cases} \tag{6.2}$$

where learning rate η_k is having the form of

$$\eta_k = \eta_0 \left[1 - \frac{\text{epoch number}}{2 \times (\text{LPC predictor size} + 1)} \right] \tag{6.3}$$

Thus, with the change of epoch number, different W_i will be obtained.

- PNN: The PNN is based on statistical principles derived from Bayes decision strategy and nonparametric kernel-based estimators of probability density function (pdf)s. It finds the pdf of features of each class from the provided training samples using Gaussian kernel. These estimated densities are then used in a Bayes decision rule to perform the classification. Advantage of the PNN is that it is guaranteed to approaches the Bayes optimal decision with a high probability. The Bayesian decision criterion is the optimal classification procedure if the pdfs of the classes to be discriminated are known *apriori*. PNN is the fastest ANN since its learning process requires only localized parameter adaptations with a single pass through the training set [8]. PNNs handle data that has spikes and points outside the norm better than other ANNs. Therefore, PNN is suitable for problems like phoneme classification. The PNN used in this work has the structure as shown in Fig. 6.1. Each pattern unit of the PNN forms a dot product of the input clean phoneme vector X with a weight vector W_i, $Z_i = X \cdot W_i$, and then performs a non-linear operation on Z_i before directing its activation level to the summation layer. Thus, the dot product is followed by a non-linear neuron activation function of the form

$$\exp[(Z_i - 1)/\sigma^2] \tag{6.4}$$

where, σ is the smoothing parameter. The summation units simply add the inputs from the pattern units that correspond to the category from which the training pattern was selected. The output, or decision, units of PNN are two-input neurons, which produces binary outputs between the two phoneme patterns obtained from the summation layer. The smoothing parameter, σ, plays a very important role in proper classification of the input phoneme patterns. Since it controls the scale factor of the exponential activation function, its value should be the same for every pattern unit. As described in [8], a small value of σ causes the estimated parent density function to have distinct modes corresponding to the locations of the training samples. A larger value of σ produces a greater degree of interpolation between points. A very large value of σ would cause the estimated density to

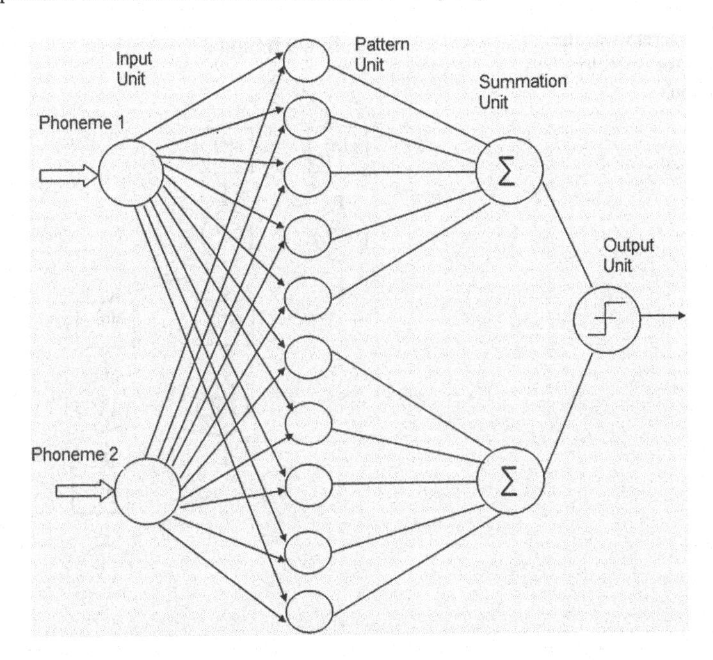

Fig. 6.1 Structure of probabilistic neural network

be Gaussian regardless of the true underlying distribution. However, it has been found that in practical problems, it is not difficult to find a good value of σ, and the misclassification rate does not change dramatically with small changes in σ.

6.4 DWT-Based Speech Segmentation

Spectral analysis is an efficient method for extracting information from speech signals. DWT has been successfully used in many speech processing applications for the spectral analysis of signals. In wavelet analysis, the signal to be analyzed is multiplied with a wavelet function, and then the transform is computed for each segment generated, where the width of the wavelet function changes with each spectral component. At high frequencies, it gives appropriate time resolution and poor frequency resolution, while at low frequencies, the DWT gives good frequency resolution and poor time resolution.

The DWT approach to speech segmentation is based on a filer bank. A discrete wavelet filter bank (DWFB) is made up of successive high-pass and low-pass orthogonal filters as shown in Fig. 6.2. The DWFB is derived from the process of wavelet decomposition tree. The timescale representation of the signal to be analyzed is passed through filters with different cutoff frequencies at different scales. Wavelets have energy concentrations in time and are useful for the analysis of transient signals

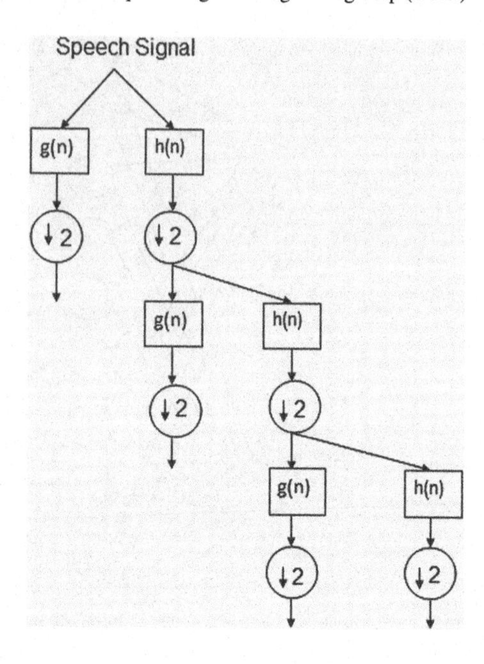

Fig. 6.2 Discrete wavelet filter bank for speech segmentation

such as speech. DWFB transforms the signal into a form that provides both the time and frequency information of the signal and is computed by successive low-pass filtering and high-pass filtering to construct a multi-resolution time–frequency plane. In DWT, to reduce the number of samples in the resultant output, a downsampling factor of $\downarrow 2$ is applied [9]. Wavelets can be realized by iteration of filters with rescaling. The resolution of the signal, which is a measure of the amount of detail information in the signal, is determined by the filtering operations, and the scale is determined by upsampling and downsampling operations [10]. Similarly, at the time of reconstruction at each level, an upsampling factor of $\uparrow 2$ is applied. Thus, the speech signal is decomposed into high- and low-frequency component and reconstructing these decomposed version of the signal, at each level, different segment of the signal can be obtained.

There are a number of basic functions that can be used as the mother wavelet for the wavelet transformation. The details of the particular application should be taken into account, and the appropriate mother wavelet should be chosen in order to use the wavelet transform effectively. Daubechies wavelets, used in this work, are the most popular wavelets that represent the foundations of wavelet signal processing and are used in numerous speech processing applications [10, 11].

In order to use this DWFB-based technique for phoneme recognition, the DWFB is modified as shown in Fig. 6.3. Here, the speech components obtained from each level of DWFB are employed for LPC feature extraction and are stored as Dw1, Dw2, Dw3, Dw4, Dw5, and Dw6. These are considered as different segments of the spoken words from which the phoneme segments are identified using the PNN-based decision algorithm explained in Sect. 6.5.3.

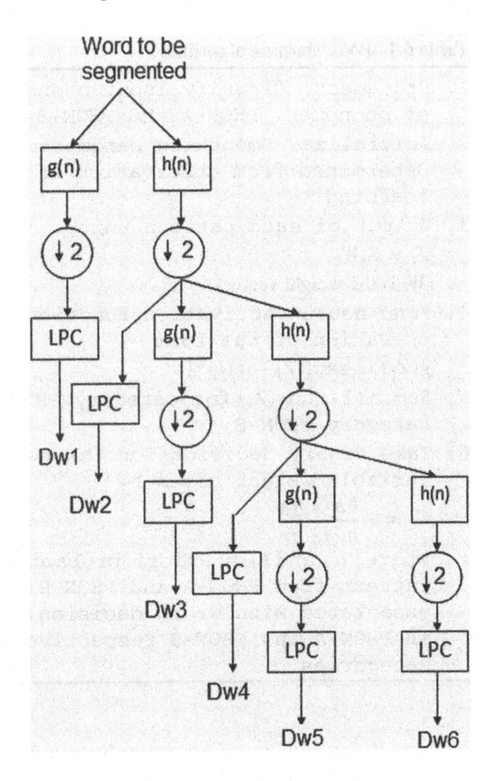

Fig. 6.3 DWT segmentation block

6.5 Proposed SOM- and PNN-Based Segmentation Algorithm

The weight vector obtained by training a one-dimensional SOM with the LPC features of a word containing the phoneme to be segmented is used to represent the phoneme. Training the same SOM with various iteration number, we get different weight vectors as explained in the Sect. 6.2, each of which is considered as a segment of different phonemes constituting the word. This chapter explains the SOM segmentation technique for segmentation of initial phoneme from CVC-type Assamese words belonging to the same environment /-at/. The weight vectors thus obtained are classified by some two-class PNNs. These PNNs are trained to learn the patterns of the initial phonemes to be segmented. At first, the work contains only the unaspirated phoneme family, which have phonemes like /p/, /b/, /t/, /d/, /k/, and /g/ that means the word which is provided for segmentation is initialized with any of these six phonemes. Three PNNs are trained with these clean unaspirated phonemes, and they are used sequentially to identify the SOM-segmented phonemes. The proposed segmentation algorithm can be stated in three distinct parts:

- PNN learning algorithm
- SOM weight vector extraction algorithm
- PNN-based decision algorithm.

Table 6.1 PNN learning algorithm

1. Statement: Classify input phoneme patterns X into two category of phonemes, PHON-A and PHON-B
2. Initialize: Smoothing parameter $\sigma = 0.111$
   Determined from observation of successful learning)
3. Output of each pattern unit,
   $Z_i = X.W_i$
   (W_i is the weight vector)
4. Find neuron activation function by performing non-linear operation of the form
   $g(Z_i) = \exp[(Z_i - 1)/\sigma^2]$
5. Sum all the$g(Z_i)$ for category PHON-A and do the same for category PHON-B
6. Take binary decision for the two summation outputs with variable weight given by-
   $C_k = -\frac{h_B I_B}{h_A I_A} \frac{n_A}{n_B}$
   where, h_A and h_B = Priori probability of occurrence of pattern from PHON-A and PHON-B respectively I = Loss associated with wrong decision n_A and n_B= No of patterns in PHON-A and PHON-B respectively which is 10 for both categories

The following sections describe each of these algorithms separately.

6.5.1 PNN Learning Algorithm

We have performed a two-class PNN problem, where three PNNs are trained with two unaspirated clean phonemes and are named as ***PNN1***, ***PNN2***, and ***PNN3***, that is, the output classes of ***PNN1*** are /p/ and /b/, the output classes of ***PNN2*** are /t/ and /d/, and the output classes of ***PNN3*** are /k/ and /g/. Clean phonemes are recorded from five boy and five girl speakers which are used as the inputs in the input layer of PNN and provided to each neuron of PNN pattern layer. The PNN learning algorithm can be stated as in Table 6.1.

The reason behind the use of two-class PNNs is nothing but to increase the success rate by performing a binary-level decision. Although it increases the computational complexity and memory requirements, it is tolerable for the sake of increasing success rate. Since PNN is the fastest, use of three PNN does not effect the speed of the algorithm.

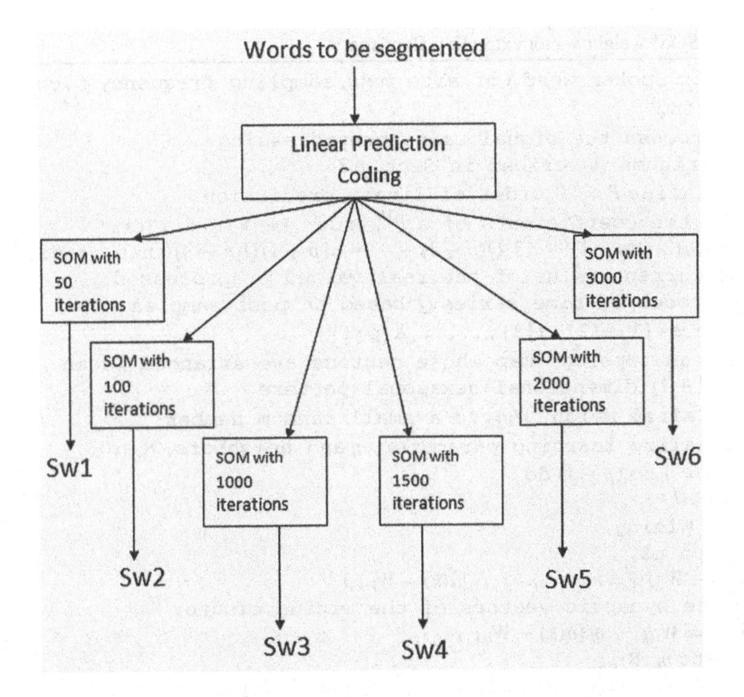

Fig. 6.4 SOM segmentation block

6.5.2 SOM Weight Vector Extraction Algorithm

The SOM weight vector extraction algorithm can be visualized from the block diagram of Fig. 6.4. The algorithm for a particular iteration can be mathematically stated as in Table 6.2. The SOM weight vector thus extracted are stored as SW1, SW2, SW3, SW4, SW5, and SW6. SOM's role is to provide segmentation boundaries for the phonemes. Here, six different segmentation boundaries are obtained for six separate sets of weight vectors. The segmented phonemes are next applied to the PNN-based decision algorithm (Sect. 6.5.3) which performs the role of decision making. Figure 6.5 shows how SOM weight vectors change with the change of iteration number.

6.5.3 PNN-Based Decision Algorithm

As mentioned earlier, the segmented phonemes are identified with the help of three PNNs trained with clean phonemes. A decision tree is designed with the three PNNs, where all the SOM or DWT provided segments are presented for pattern matching one by one. The algorithm for a particular segment X can be stated as in Table 6.3.

Table 6.2 SOM weight vector extraction algorithm

```
1. Input: Spoken word S of size m×n, sampling frequency f_s, duration
   T second
2. Preprocess the signal using preprocessing
   algorithms described in Sect. 6.2
3. Initialize P = 20, order of linear prediction
4. Find the coefficients of a P^th-order LP FIR filter,
   Q̂ = -a(2)Q(n-1) - a(3)Q(n-2) - ··· - a(p+1)Q(n-p) that predicts
   the current value of the real-valued preprocessed
   preprocessed time series Q, based on past samples
4. Store a=[1,a(2),a(3),.....,a(p+1)]
5. Take an topology map whose neurons are arranged in an
   1 × (P+1)-dimensional hexagonal pattern
6. Initialize weight W_i(k) to a small random number
7. Initialize learning parameter, η and neighbors, ℵ_I((k)
8. for k = 1 to (p+1) do
   pick a(k)
   find Wining
   neuron as,
   |a(k) - W_i(k)| = min_(1≤i≤(p+1))|a(k) - W_i(k)|
   Update synaptic vectors of the wining cluster,
   W_i(k) = W_i(k) + η_k(a(k) - W_i(k))
   Update η_k, ℵ_I((k)
9. Store updated weight W_i(k) as SW_j, where j=1:6
```

6.6 Experimental Details and Result

The process logic of the SOM-based phoneme segmentation technique is shown in Fig. 6.6. All the speech signals are first preprocessed using the preprocessing algorithm described in Sect. 6.6.2, and then LPC coefficients are extracted.

6.6.1 Experimental Speech Signals

The speech sample database used in the proposed segmentation technique can be explained as follows:

At the first phase, clean unaspirated phoneme segments $/p/$, $/b/$, $/t/$, $/d/$, $/k/$, and $/g/$ are generated by manual marking from a set of words recorded from five girl speakers and five boy speakers to train the PNNs. A word list is prepared for the work as shown in Table 6.4 which covers words having initial phoneme variation between all the unaspirated phonemes. All these words are recorded from five girl and five boy speakers, which are from various part of the state of Assam, and therefore, these possess variations in pronunciation. This variation is considerable. There are six unaspirated phonemes. So with ten speakers, it turns out to be sixty

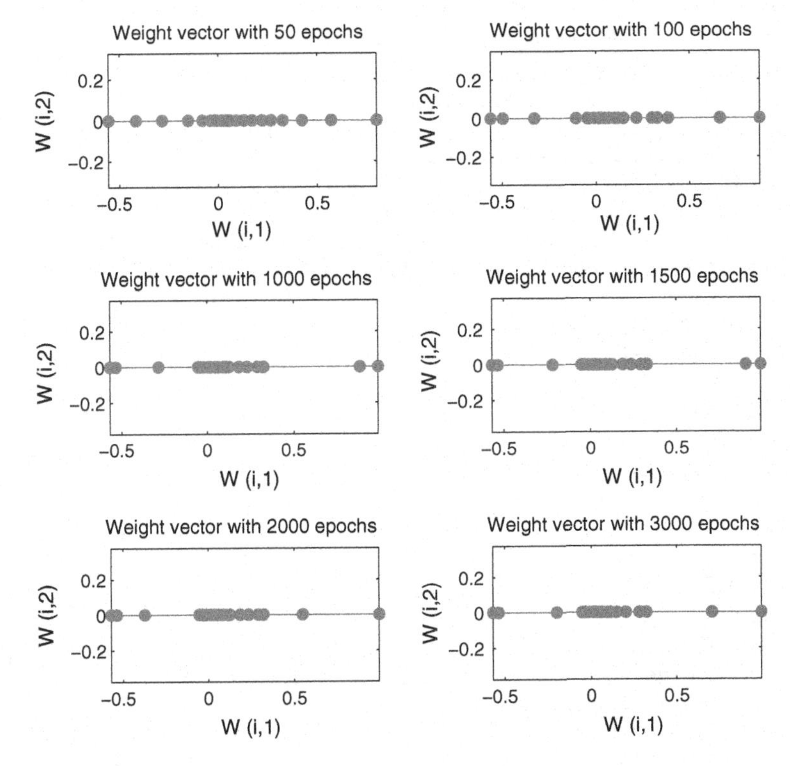

Fig. 6.5 SOM weight vectors

segmented samples. Another set of samples are recorded from some other speakers which are used for validation and testing of PNNs. This is the size of data used with the work. This database was created to serve the purpose of the work. For recording the speech signal, a PC headset and a sound-recording software, Gold Wave, is used. The recorded speech sample has the following specification:

- Duration = 2 s,
- Sampling rate = 8000 samples/s, and
- Bit resolution = 16 bits/sample.

6.6.2 Preprocessing

The preprocessing of the speech signal consists of two operations—smoothing of the signal by median filtering and removal of the silent part by threshold method. Although the speech signals are recorded in a noise-free environment, presence of some unwanted spikes are observed. Therefore, a median filtering operation is

Table 6.3 PNN-based decision algorithm for consonant phoneme

```
1. Input: Speech S of size m × n, sampling
   frequency fs, duration T second
2. Preprocess the signal using preprocessing
   algorithms described in Sect. 6.2
3. Obtain DW1, DW2, DW3, DW4, DW5, and DW6 using the DWT
   based segmentation described in Sect. 6.4
4. Obtain SW1, SW2, SW3, SW4, SW5, and SW6 using the SOM weight
   vector extraction algorithm described in Sect. 6.5.2
5. Load PNN1
6. Decide PHON-A
   If DW1=PHON-A
   else if
   DW2= PHON-A
   else if
   DW3= PHON-A
   else if
   DW4=PHON-A
   else if
   DW5=PHON-A
   else if
   DW6=PHON-A
   else Decide
   'Not Assamese Phoneme /X/'.
7. or Decide PHON-A
   If SW1=PHON-A
    else if
   SW2=PHON-A
   else if    SW3= PHON-A
    else if
   SW4=PHON-A     else if
   SW5=PHON-A
   else if
   SW6=PHON-A
   else Decide
   'Not Assamese Phoneme /X/'.
```

performed on the raw speech signals, so that the vowel segmentation does not suffer by any types of unwanted frequency component [12, 13].

The smoothed signal S_{smooth} contains both speech and non-speech part. The non-speech or silent part occurs in a speech signal due to the time spend by the speaker before and after uttering the speech, and this time information is considered to be redundant for phoneme segmentation purpose. The silent part ideally have zero intensity. But, in practical cases, it is observed that even after smoothing, the silent part of the speech signal has intensity about 0.02. Our silent-removing algorithm considers this intensity value as a threshold. Thus, a pure signal containing only the

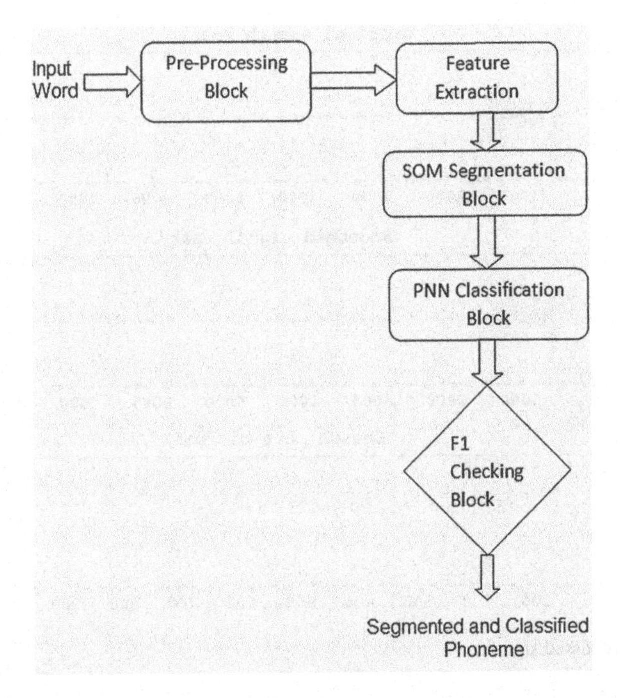

Fig. 6.6 Process logic of the SOM-based segmentation algorithm

Table 6.4 Word list with unaspirated phoneme at word initial position

Sl No	Unaspirated phoneme	Word
1	/p/	/pat/
2	/b/	/bat/
3	/t/	/tat/
4	/d/	/dat/
5	/k/	/kat/
6	/g/	/gat/

necessary speech part is obtained, as shown in Fig. 6.7. Tables 6.5 and 6.6 show the smoothing and silent-removing algorithm.

6.6.3 Role of PNN Smoothing Parameter

Initially, three PNNs are trained with manually marked /p/, /b/, /t/, /d/, /k/, and /g/ phonemes. Before using these PNNs for identification of segmented phonemes, it is tested with similar manually marked phoneme segments. The smoothing parameter, σ, plays a very important role in proper classification of the input phoneme patterns during training. Since it controls the scale factor of the exponential activation

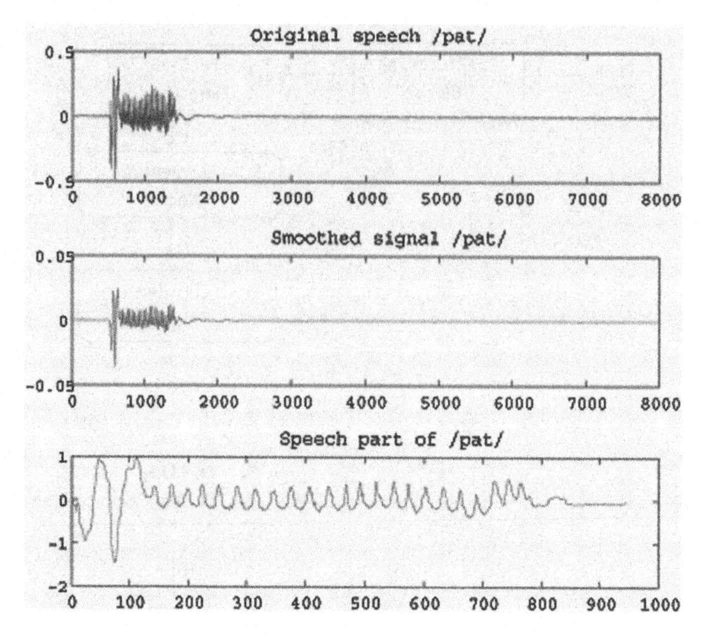

Fig. 6.7 Preprocessed signal

Table 6.5 Smoothing algorithm

1. Input: Speech S of size m× n, sampling frequency f_s, duration T second
2. Output: Smoothed signal S_{smooth} of size 1× n
3. for $i = 1$ to n do for $j = 1$ to m do
4. $y_i = (S_i + S_{j+1})/2$
5. end for
6. end for
7. initialize $J = 2$
8. S_{smooth}=median$[y(i-(J/2)), y(i-(J/2)+1), \ldots, y(i+(J/2)-1)]$

function, its value should be the same for every pattern unit. As described in [8], a small value of σ causes the estimated parent density function to have distinct modes corresponding to the locations of the training samples. A larger value of σ produces a greater degree of interpolation between points. A very large value of σ would cause the estimated density to be Gaussian regardless of the true underlying distribution. However, it has been found that in practical problems, it is not difficult to find a good value of σ, and the misclassification rate does not change dramatically with small changes in σ. Table 6.7 shows the rate of correct classification for a range of σ value, and it can be seen that from 0.100 if we start to decrease, PNN shows 100 % correct classification, and therefore, PNN trained with these values are found to be useful for testing. For testing, a decision tree is designed using the three PNNs, where all the

Table 6.6 Silent-removing algorithm

1.Input: Smoothed signal S_{smooth} of size $1 \times n$, duration T second
2.Output: The speech part of the signal Q of new size $1 \times q_{size}$,
3.Break the signal into frames of 0.1 seconds
  Initialize frame duration $= 0.1$
  Initialize threshold $= 0.03$
  $Framelength = frameduration \times fs$
  $N = length(xin)$
  $numberofframes = floor(N/frame_len)$
  $Q = zeros(N, 1)$
  $count = 0$
  $fork = 1 : numofframes$
4.Extract a frame of speech
  $frame = S((k-1) \times Framelength + 1 : Framelength \times k)$
5.Identify non-silent frames by finding frames with max amplitude more than a threshold say 0.03
  $maxvalue = maximum\ of\ frame$
  if
  $maxvalue > threshold$
  $This\ frame\ is\ not\ silent$
  $count = count + 1$
  $Q((count-1) \times frame_len + 1 : frame_len \times count) = frame$
  end
  end
6.Detect the starting point of unwanted zeros in Q.
  (These zeros come due to the initialization of the zero matrix for Q.)
  $fori = 1 : N$
  if $Q(i) == 0$
  $x = i$
  $break$
  end
  end
  $Q(x : end) = []$

PNNs are checked one by one. The recognition rate of the four PNNs with $\sigma = 0.111$ while tested with manually marked vowel are shown in Table 6.8. The experiments are repeated for several trials, and the success rates are calculated. After obtaining a satisfactory success rate for manually marked phonemes, the same PNN decision tree is used to identify phoneme segment provided by SOM or DWT.

Table 6.7 PNN classification rate for various σ value

Sl No	σ	Correct classification (%)	Misclassification (%)
1	0.0001	100	0
2	0.001	100	0
3	0.01	100	0
4	0.1	100	0
5	0.111	100	0
6	0.222	100	0
7	0.333	93.75	6.25
8	0.444	93.75	6.25
9	0.555	93.75	6.25
10	0.666	87.5	12.25
11	0.777	81.25	18.75
12	0.888	81.25	18.75
13	0.888	81.25	18.75
14	0.999	81.25	18.75
15	1	81.25	18.75
16	2	75	25
16	2	75	25
17	10	75	25
18	30	75	25
16	50	75	25
16	100	75	25

Table 6.8 Success rate of PNN while tested with manually segmented phonemes

Sl No	Vowel	Correct recognition (%)	False acceptance (%)	Unidentified (%)
1	/p/	92	4.4	4.6
2	/b/	97	4	1
3	/t/	98.2	1.2	0.6
4	/d/	94	4	2
5	/k/	94	1.8	3.2
6	/g/	97	3	0

6.6.4 Comparison of SOM- and DWT-Based Segmentation

As mentioned earlier, the segmentation is carried out twice by two different methods. First, using conventional DWT-based method and secondly using a SOM-based method. The segmented phonemes are then checked one by one to find which particular segment represents the initial phoneme by matching pattern with the trained PNNs. It is observed that the recognition success rate is abruptly increased (from 89.7 to 97 %) with the SOM-based segmentation technique. Table 6.9 summarizes this performance difference.

Table 6.9 DWT versus SOM segmentation success rate for initial phoneme segmentation

Sl No	Segmentation technique	Success rate (%)
1	DWT	89.3
2	SOM	96

Table 6.10 DWT versus SOM segmentation success rate for unaspirated phonemes in word initial position

Sl No	Phoneme	Success rate of DWT (%)	Success rate of SOM (%)
1	/p/	90	98
2	/b/	86	95
3	/t/	85	93
3	/d/	85	90
3	/k/	95	99
3	/g/	95	99

Table 6.11 DWT versus SOM segmentation computational time

Sl No	Segmentation technique	Correct decision (s)	Wrong decision (s)
1	DWT	6–7	9–10
2	SOM	50–60	70–80

The SOM is trained for six different iterations, which provides identical number of decision boundaries. For 50 sessions, a segment named SW_1 is obtained (Fig. 6.4). Similarly, for 100, 1000, 1500, 2000, and 3000 sessions, other five segments are obtained. This is somewhat similar to the segmentation carried out using DWT. The DWT provides six levels of decomposition with faster processing time than the SOM (Table 6.11). In case of SOM, some vital time slots are lost in training. But, the segments provided by SOM provide better resolution which helps the PNN to provide improved discrimination capability subsequently. SOM with its ability to reorganize can capture context better and thereby help in proper segmentation of phoneme boundaries. Its ability to generate different topological states with varying epochs enables it to capture contextual variations better with changing number of epochs. Table 6.10 shows success rate for various phonemes.

The disadvantage of the proposed SOM-based algorithm is that the computational complexity is more compared to the DWT-based technique, which in turn increases the computational time. The computational complexity of the SOM-based algorithm can be analyzed as

```
Suppose, N be the size of the input pattern which is one added to
the LP predictor length. This value is 21.
Here, N=LP Predictor Length +1= 21,
Now, for a single epoch,
1st step, find the wining neuron:
Number of summation= 21,
2nd step, update synaptic vectors of the wining cluster:
Number of summation= 42,
```

```
Number of multiplication= 42,
3rd step, update learning parameter:
Number of summation= 2,
Number of multiplication= 3,
Thus, computational complexity for a single epoch,
Total number of summations= 65
number of multiplications= 45
Minimum computational complexity for this algorithm, that is, for
50 epoch,
number of summation= 50× 65=154.7N
number of multiplication= 50× 45=107.1N
Maximum computational complexity for this algorithm, that is, for
3000 epoch
number of summation= 3000× 65=9285.7N
number of multiplication= 3000×45=6428.5N
```

Compared to it, DWT has maximum computational complexities of 14.18 N and 24.24 N. Since success rate reaches a satisfactory level and precision is much improved compared to DWT, increase in speed is tolerable. Table 6.11 shows variation in computational time with DWT and SOM algorithm. It is to be noted that if the decision is wrong, then more time is taken since the algorithm passed through each and every decision level. The computational complexity of the present approach can be reduced by a number of ways like better hardware, better programming approaches, concurrent training of the SOMs, including distributed processing environments, and use of specialized DSP processors. Presently, it has been implemented on a 2007 model Intel dual core P4 machine with 2 GB RAM.

6.7 Conclusion

The work describes the design of an efficient phoneme segmentation technique. A sample dataset is prepared for the work considering the unaspirated phonemes in word initial position of the language. Here, we have proposed a new segmentation algorithm based on SOM and PNN with a prior knowledge of formant frequency of Assamese phonemes. The algorithm shows distinct advantage in terms of segmentation success rate over conventional DWT-based segmentation technique. Although the computational time increases, it is tolerable for the sake of increasing success rate, and it can be reduced in future by reducing the computational complexity. Further, by using a higher-level decision strategy to move the algorithm control from one phoneme family to the other, we can extend the algorithm to recognize spoken word of Assamese language. Such an application is described in Chap. 7.

References

1. Grayden DB, Scordilis MS (1994) Phonemic segmentation of fluent speech. Proc IEEE Int Conf Acoust, Speech, Signal Process 1:73–76
2. Sweldens W, Deng B, Jawerth B, Peters G (1993) Wavelet probing for compression based segmentation. Proc SPIE 2034:266–276
3. Wendt C, Petropulu AP (1996) Pitch determination and speech segmentation using the discrete wavelet transform. IEEE In Symp Circuits Syst Connect World 2:45–48
4. Zioko B, Manandhar S, Wilson RC (2006) Phoneme segmentation of speech. Proc 18th Int Conf Pattern Recogn 4:282–285
5. Paliwal KK, Kleijn WB (2003) Quantization of LPC parameters. J Inf Sci Eng 19:267–282
6. Kohonen T (1990) The self-organizing map. Proc IEEE 78(9):1464–1480
7. Haykin S (2009) Neural network and learning machine, 3rd edn. PHI Learning Private Limited, India
8. Specht DF (1990) Probabilistic neural networks. Neural Netw 3(109):118
9. Shah FA, Sukumar RA, Anto BP (2010) Discrete Wavelet transforms and artificial neural networks for speech emotion recognition. Int J Comput Theor Eng 2(3)
10. Sripathi D (2003) Chapter 2, The discrete wavelet transform. Available at http://etd.lib.fsu.edu/theses/available/etd-11242003-185039/unrestricted/09dschapter2.pdf
11. Long CL, Datta S (1996) Wavelet based feature extraction for phoneme recognition. In: Proceedings of 4h international conference on spoken language, vol 1
12. Rabiner LR, Schafer RW (2009) Digital processing of speech signals. Pearson Education, Dorling Kindersley (India) Pvt Ltd, India
13. Tang BT, Lang R, Schroder H (1994) Applying wavelet analysis to speech segmentation and classification. Wavelet applications. In: Proceedings of SPIE, 2242

Chapter 7
Application of Phoneme Segmentation Technique in Spoken Word Recognition

Abstract This chapter explores the application possibility of the SOM-based phoneme segmentation technique in the field of isolated spoken word recognition. A hybrid framework is designed using RNN, PNN, and LVQ to recognize consonant–vowel–consonant (CVC)-type Assamese words.

Keywords Linear prediction coding · Formant frequency · SOM · PNN · RNN · LVQ

7.1 Introduction

Spoken word recognition is a distinct subsystem providing the interface between low-level perception and cognitive processes of retrieval, parsing, and interpretation of speech. The process of recognizing a spoken word starts from a string of phonemes, establishes how these phonemes should be grouped to form words, and passes these words onto the next level of processing. As discussed in Chap. 2, there are several theories in the literature, which focuses on the discrimination and categorization of speech sounds. One of the earliest theories in this contrast is the motor theory proposed by Alvin Liberman, Franklin Cooper, Pierre Delattre, and other researchers in 1950. Motor theory postulates that speech is perceived by reference to how it is produced. It means that when perceiving speech, listeners access their own knowledge of how phonemes are articulated. Analysis by synthesis model, proposed by Stevens and Halle, 1960, in turn stated that speech perception is based on auditory matching mediated through speech production. Research on the discrimination and categorization of phonetic segments was the key focus of the works on speech perception before 1970s. The processes and representations responsible for the perception of spoken words became a primary object of scientific inquiry with a curiosity of disclosing the cause and methods of how the listener perceives fluent speech. According to Cohort theory (1980) [1, 2], various language sources such as lexical, syntactic, and

M. Sarma and K. K. Sarma, *Phoneme-Based Speech Segmentation Using Hybrid Soft Computing Framework*, Studies in Computational Intelligence 550,
DOI: 10.1007/978-81-322-1862-3_7, © Springer India 2014

semantic interact with each other in complex ways to produce an efficient analysis of spoken language. It suggests that inputs in the form of a spoken word activates a set of similar items in the memory, which is called word initial cohort. The word initial cohort consisted of all words known to the listener that begin with the initial segment or segments of the input words. In 1994, Marslen-Wilson and Warren revised the cohort theory. In the original version, words were either in or out of the word cohort. But in the revised version, word candidates vary in the activation level, and so membership of the word initial cohort plays a important role in the recognition [1]. According to the TRACE model of spoken word recognition, proposed by McClelland and Elman, in 1986, speech perception can be described as a process, in which speech units are arranged into levels which interact with each other. There are three levels: features, phonemes, and words. There are individual processing units or nodes at three different levels. Dennis Norris in 1994 have proposed another model of spoken word recognition called the shortlist model. According to Norris, a short list of word candidates is derived at the first stage of the model. The list consists of lexical items that match the bottom-up speech input. This abbreviated list of lexical items enters into a network of word units in the later stage, where lexical units compete with one another via lateral inhibitory links [2]. Thus from this discussion, we arrived at the conclusion that the spoken word recognition is a complex multileveled pattern recognition work performed by neural networks of human brain, and speech perception process can be modeled as a pattern recognition network. Different levels of the language such as lexical, semantic, and phonemes can be used as the unit of distribution in the model. Most of the spoken word recognition model provides the same idea that various speech units are activates at different level of recognition, and decisions at various level are combined to take a global decision about the spoken word. This work has proposed a spoken word recognition model, where a set of word candidates are activated at first on the basis of phoneme family to which its initial phoneme belongs. The phonemical structure of every natural language provides some phonemical groups for both vowel and consonant phonemes each having distinctive features. This work provides an approach to consonant–vowel–consonant (CVC)-type Assamese spoken words recognition by taking advantages of such phonemical groups of Assamese language, where all words of the recognition vocabulary are initially classified into six distinct phoneme families, and then, the constituent vowel and consonant phonemes are classified within the group. A hybrid framework, using four different ANN structures, is constituted for this word recognition model, to recognize phoneme family and phonemes and thus the word at various level of the algorithm. Recognition of individual phoneme from a word must need some efficient segmentation algorithm which can extract the differentiating characteristics of that particular phoneme, and it ultimately leads to the recognition of the word. The work uses the proposed SOM-based approach to segment and recognize the vowel and consonant phonemes from some two alphabet CVC-type Assamese words, where the SOM segments the word into its constituent phonemes in an unsupervised manner, whereas two supervised ANN structures, RNN and PNN, respectively, are playing the role of recognizing initial phoneme from the phoneme segments. Some two class PNNs are trained to learn the patterns of all the consonant phonemes of Assamese language. The phoneme segments obtained by training SOM with various iteration numbers is then matched with the PNN patterns. While taking decision about the last phoneme, the algorithm is assisted by LVQ codebook which contains a distinct code for every word of the recognition vocabulary. Before the global decision taken by the PNN, a RNN takes some local decisions about the incoming word and classifies them into six phoneme families of Assamese language. The segmentation and recognition are performed separately at the RNN decided family. An important aspect of the classification algorithm performed by PNN is that it uses the priori knowledge of first formant frequency (F1) of the Assamese phonemes while taking decisions about the vowel phoneme, since vowels are distinguished by its own unique pattern as well as in terms of their formant frequencies. Here, the concept of pole or formant location determination from the linear prediction model of vocal tract used while estimating F1 for all the Assamese vowels. Assamese phonemes and words containing those phonemes are recorded from five girls and five boys, so that the recognition algorithm can remove the speaker dependance limitation.

7.2 Linear Prediction Model for Estimation of Formant Frequency

Speech signals are produced from the air forced through the vocal cords by the diaphragm. Vocal tract, which acts as a resonator, modifies the excitation signal causing formant or pole and sometimes antiformant or zero frequencies [3]. In general, the fundamental frequency and formant frequencies are the most important concepts in speech processing. Speech can be represented phonetically by a finite set of symbols called the phonemes of the language, the number of which depends upon the language and the refinement of the analysis. For most languages, the number of phonemes lies between 32 and 64. Each phoneme is distinguished by its own unique pattern, and different phoneme is distinguishable in terms of their formant frequencies.

As part of this work, the formant frequencies of Assamese vowel sounds are estimated which are used as a prior knowledge in the proposed SOM-based segmentation technique. The linear prediction model of speech signal, which gives the best spectral model of the vocal tract, is used in the technique. Vowel sounds from five girls and five boys are recorded for the work so that the result can reflect a proper estimation of formant frequencies of Assamese vowels. The estimation of the formant frequencies, mainly the first three formants, F1, F2, and F3 has many practical applications. They are used for the characterization of the different sounds found in the speech [4].

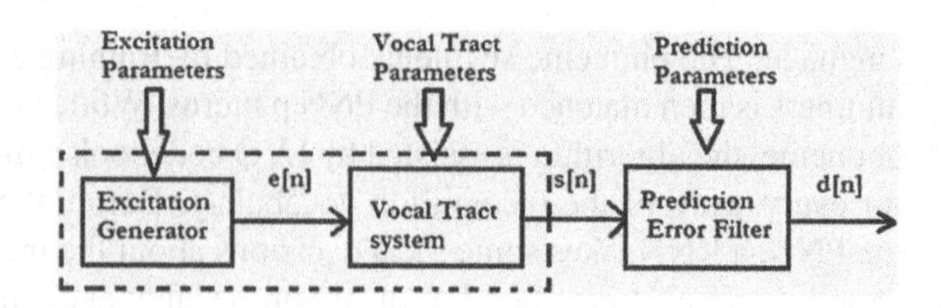

Fig. 7.1 Source filter model of vocal tract with prediction filter

7.2.1 Human Vocal Tract and Linear Prediction Model of Speech

The speech signal is produced by the action of the vocal tract over the excitation coming from the glottis. Different conformations of the vocal tract produce different resonances that amplify frequency components of the excitation, resulting the different sounds [3]. The sounds created in the vocal tract are shaped in the frequency domain by the frequency response of the vocal tract. The resonance frequencies resulting from a particular configuration of the articulators are instrumental in forming the sound corresponding to a given phoneme. These resonance frequencies are called the formant frequencies of the sound. The resonances of the vocal tract tube shape the sound sources into the phonemes. The source filter model of the vocal tract system as in Fig. 7.1 can be represented by discrete time linear time invariant filter. The short-time frequency response of the linear system simulates the frequency shaping of the vocal tract system, and since the vocal tract changes shape relatively slowly, it is reasonable to assume that the linear system response does not vary over time intervals on the order of 10 ms or so. Thus, it is common to characterize the discrete time linear system by a system function of the form given in Eq. 7.1

$$\begin{aligned} H(z) &= \frac{\sum_{k=0}^{M} b_k Z^{-k}}{1 - \sum_{k=1}^{N} a_k Z^{-k}} \\ &= \frac{b_0 \prod_{k=1}^{M}(1 - d_k z^{-1})}{\prod_{k=1}^{N}(1 - c_k z^{-1})} \end{aligned} \tag{7.1}$$

where the filter coefficients a_k and b_k change at a rate on the order of 50–100 times/s. Those poles c_k of the system function lie close to the unit circle and create resonances to model the formant frequencies [5].

The linear predictive model is based on a mathematical approximation of the vocal tract. At a particular time, t, the speech sample $s(t)$ is represented as a linear sum of the p previous samples. In LP model of speech, each complex pole pair corresponds to a second-order resonator. The resonance frequency of each pole is associated with a peak in spectral energy or a formant candidate. The pole radius is related to the concentration of local energy and the bandwidth of a formant candidate.

Over short-time intervals, the linear system as given by Eq. 7.1 can be described by an all-pole system function of the form:

$$H(Z) = \frac{S(Z)}{E(Z)} = \frac{G}{1 - \sum_{k=1}^{P} a_k Z^{-k}} \tag{7.2}$$

The major advantage of this model is that the gain parameter, G, and the filter coefficients a_k can be estimated in a very straightforward and computationally efficient manner by the method of linear predictive analysis.

In linear prediction, the speech samples $s[n]$ are related to the excitation $e[n]$ by the difference equation given in Eq. 7.3

$$s[n] = \sum_{k=1}^{P} a_k s[n-k] + Ge[n] \tag{7.3}$$

A linear predictor with prediction coefficients, α_k, is defined as a system whose output is given by Eq. 7.4

$$\tilde{s}[n] = \sum_{k=1}^{P} \alpha_k s[n-k] \tag{7.4}$$

and the prediction error, defined as the amount by which $\tilde{s}[n]$ fails to exactly predict sample $s[n]$, is

$$d[n] = s[n] - \tilde{s}[n] = s[n] - \sum_{k=1}^{P} \alpha_k s[n-k] \tag{7.5}$$

The linear prediction error sequence possesses the form of the output of an FIR linear system whose system function is

$$A(Z) = 1 - \sum_{k=1}^{P} \alpha_k Z^{-k} \tag{7.6}$$

The system function of the prediction error filter is a polynomial in z^{-1}, and therefore, it can be represented in terms of its zeros as

$$A(Z) = \prod_{k=1}^{P} (1 - Z_k Z^{-1}) \tag{7.7}$$

Thus, the prediction error filter, $A(z)$, is an inverse filter for the system, $H(z)$, i.e.,

$$H(Z) = \frac{G}{A(Z)} \tag{7.8}$$

According to Eq. 7.8, the zeros of $A(z)$ are the poles of $H(z)$. Therefore, if the model order is chosen judiciously, then it can be expected that roughly $fs/1{,}000$ of the roots will be close in frequency to the formant frequencies [5].

7.2.2 Pole or Formant Location Determination

The pole location from the prediction polynomial $A(Z)$ can be calculated by solving for the roots of the equation $A(Z) = 0$. Each pair of complex roots is used to calculate the corresponding formant frequency and bandwidth. The complex root pairs $Z = r_0 e^{\pm\theta_0}$ and sampling frequency f_s to formant frequency F and 3-db bandwidth B possess the transformation of the form given by the following pair of equations.

$$F = \frac{fs}{2\pi}\theta_0 Hz \tag{7.9}$$

$$B = \frac{fs}{\pi}\ln r_0 Hz \tag{7.10}$$

7.3 LVQ and its Application to Codebook Design

LVQ is a supervised version of vector quantization that can be used to create a codebook for application as a classifier. It models the discrimination function defined by the set of labeled codebook vectors and the nearest neighborhood search between the codebook and data. An LVQ network has a first competitive layer and a second linear layer. The competitive layer learns to classify input vectors in much the same way as the competitive layers of SOM. The linear layer transforms the competitive layer's classes into target classifications defined by the user. The classes learned by the competitive layer are referred to as subclasses and the classes of the linear layer as target classes [6]. The training algorithm involves an iterative gradient update of the winner unit [7]. Assume that a number of codebook vectors, m_i (free parameter vectors) are placed into the input space to approximate various domains of the input vector x by their quantized values. Usually, several codebook vectors are assigned to each class of x values. The input vector x is then decided to belong to the same class to which the nearest m_i belongs [7]. Let

$$C = \arg\min_i \| x - m_i \| \tag{7.11}$$

define the nearest m_i to x, denoted by m_c.

Values for the m_i that approximately minimize the misclassification errors in the above nearest-neighbor classification can be found as asymptotic values in the following learning process. Let $x(t)$ be a sample of input and let the $m_i(t)$ represent sequences of the m_i in different time slots. Starting with properly defined initial values, the following equations define the basic LVQ1 process obtained from [7].

$$\begin{gathered}
m_c(t+1) = m_c(t) + \alpha(t)[x(t) - m_c(t)] \\
\textit{if } x \text{ and } m_c \text{ belong to the same class} \\
m_c(t+1) = m_c(t) - \alpha(t)[x(t) - m_c(t)] \\
\textit{if } x \text{ and } m_c \text{ belong to the different class} \\
m_i(t+1) = m_i(t)
\end{gathered} \tag{7.12}$$

Here, $0 < \alpha(t) < 1$ and $\alpha(t)$ may be constant or decrease monotonically with time. In the above basic LVQ1, it is recommended that α should initially be smaller than 0.1. The new factor in LVQ is that the input data points have associated class information. This allows the use of prior classification labels of the speaker source information to find the best classification label for each weight vector, i.e., for each Voronoi cell in case of vector quantization. Then during pattern matching for identification, each segmented vowel without a class label can be assigned to the class of the Voronoi cell it falls within.

7.4 Phoneme Segmentation for Spoken Word Recognition

Initial phoneme used to activate words starting with that phoneme in most of the renowned spoken word recognition model such as Cohort (1976), Trace (1986) etc [1, 8]. The speech can be firstly classified into a phonetic group, and then, it is classified within the group. Phoneme groups are categorized based on articulatory information. In this section, a new word recognition strategy is proposed to recognize CVC-type Assamese spoken word, where activation of a candidate word occurs at the first step of phoneme recognition in terms the phoneme family to which its initial phoneme belong. Then, the candidate word is segmented into its various phoneme boundary within that family, and all the vowel and consonant phonemes constituting the word are recognized. The algorithm can be summarized as in the flow diagram of Fig. 7.2. The algorithm considers words starting with all the phonemes belongs to each of the Assamese phoneme family, except /η/, /w/, and /j/ since no words in Assamese start with these phonemes [9]. According to [9], these phonemes are only used in the middle of the word. The proposed word recognition method can be stated in four distinct part:

- RNN-based local classification
- SOM-based segmentation
- PNN- and F1-based initial phoneme and vowel recognition and
- LVQ codebook assisted last phoneme recognition

The following sections describe each of these part separately.

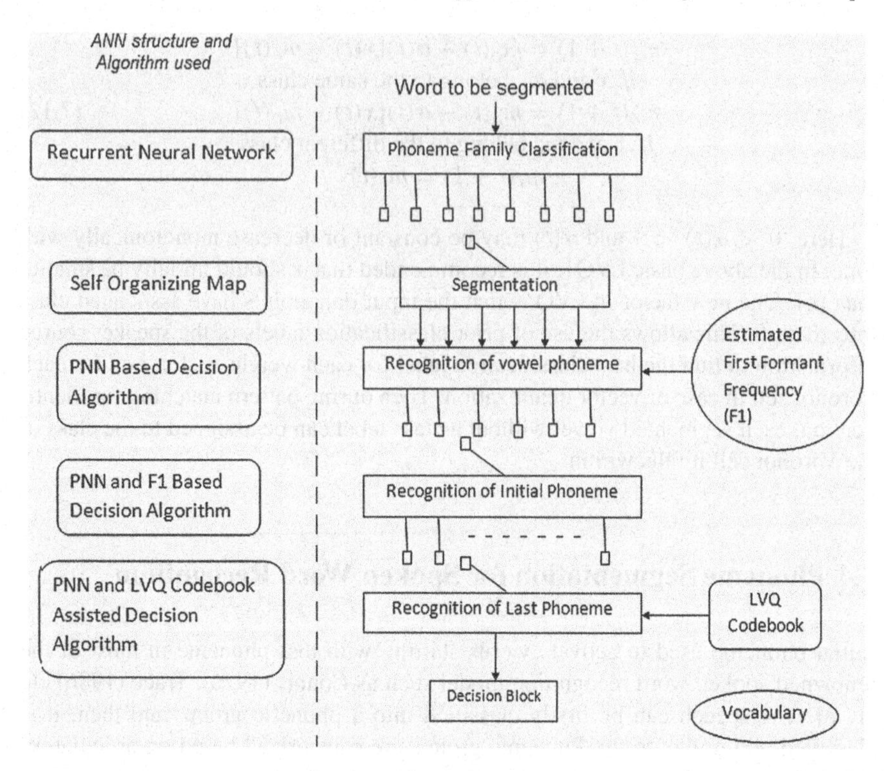

Fig. 7.2 Process logic of the spoken word recognition model

7.4.1 RNN-Based Local Classification

ANNs being nonparametric prediction tools based on the ability to learning are also suitable for use for a host of pattern classification/application including speech recognition. One of the most commonly used ANN structures for pattern classification is the MLP which have found application in a host of work related to speech synthesis and recognition. The MLPs work well as an effective classifier for vowel sounds with stationary spectra, while their phoneme discriminating power deteriorates for consonants characterized by variations of short-time spectra [10]. Feed forward MLPs are unable to deal with time-varying information as seen in the speech spectra. The RNNs have the ability to deal with time-varying nature of the inputs for which these have proven to be suitable for application like speech recognition. A few such works are as already been discussed in Sect. 1.3.1. Here, a RNN is used to classify every word of the vocabulary into the six phoneme families of Assamese on the basis of the formant frequency of its initial phoneme. The RNN will check the first formant frequency F1 of the incoming word and take a decision about the phoneme family to which the word may belong. The training carried out beforehand enables the RNN to learn the patterns and associate with them the appropriate class levels. The training quality ensures satisfactory performance during subsequent stage. The training quality ensures satisfactory performance during subsequent stage. The learning acquired by the RNN enables it to take recognition decisions. Thus, the algorithm control can move from one phoneme family to other phoneme family. In this execution, critical support is received from the segmentation process which separates out individual phonemes from an applied speech samples. The SOM-based segmentation algorithm described in the Chap. 6 is used in every classified phoneme family to segment the word into phonemes. The work is carried out by extracting LPC feature vectors of the incoming word. Before classification, the inputs are checked for F1 range and accordingly the RNN decides to which family the word may belong. Every word of the recognition vocabulary whose initial phoneme lies in the same phoneme family shows some similarity in the F1 value. Therefore, the first phoneme F1 range for all the Assamese phoneme family is estimated a priori. Thus, the range of formant frequencies are used as a priori knowledge for classification.

7.4.2 SOM-Based Segmentation Algorithm

The SOM-based segmentation technique is explained in detail in Sects. 6.3 and 6.5. The weight vectors obtained by training a one-dimensional SOM with the LPC features of a word are used to represent various phoneme segment. By training the same SOM with various iteration numbers, we get different weight vectors, each of which is considered to be a segment of the incoming word categorized as individual phonemes. These phonemes in totality constitute the word. The weight vectors thus obtained are classified according to the Assamese phonemical structure which consisted of six phoneme families as described in Chap. 4. The SOM weight vector thus extracted is stored as SW1, SW2, SW3, SW4, SW5, SW6 for subsequent processing. SOM's role is to provide segmentation boundaries for the phonemes. Here, six different segmentation boundaries are obtained clear demarkation of which identifies six separate sets of weight vectors. The segmented phonemes are next applied to the PNN-based recognition algorithm which performs the role of decision making.

7.4.3 PNN- and F1-Based Vowel Phoneme and Initial Phoneme Recognition

PNNs trained with clean vowel and consonant phonemes are used sequentially to identify the SOM segmented phonemes. Here, we have performed a two class PNN-based recognition problem, where one particular PNN is trained with two clean phonemes and is named as ***PNN1***, ***PNN2***, ***PNN3***, ***PNN4*** etc. It means that the output classes of ***PNN1*** are /p/ and /b/, the output classes of ***PNN2*** are /t/ and /d/, the output classes of ***PNN3*** are /k/ and /g/, the output classes of ***PNN4*** are /ph/ and /bh/, the output classes of ***PNN3*** are /th/ and /dh/ etc., and this way all the consonant phonemes are considered. Similarly, for vowel phonemes, also PNNs are trained like ***PNN11*** is trained with /i/ and /u/, ***PNN12*** is trained with /e/ and /o/ and so on. Thus, we obtain a total ten PNNs for consonant phonemes and four PNNs for vowel phonemes, which are considered to know patterns of all the phonemes of Assamese speech. Clean phonemes are recorded from five boy and five girl speakers which are used as the inputs in the input layer of PNN and provided to each neuron of PNN pattern layer. The PNN learning algorithm has already been stated as in Sect. 6.5.1. For initial consonant, the decision algorithm is same as given in Sect. 6.5.3. A two-step decision is taken by the decision algorithm in case of vowel phonemes. First matching the vowel segment with the PNN patterns and then checking its first formant frequency F1, whether it lies in the predetermined range or not is performed by an algorithm as summarized in Table 7.1. This is applied for vowel phonemes.

7.4.4 LVQ Codebook Assisted Last Phoneme Recognition

An LVQ codebook is used along with the PNN- and F1-based recognition algorithm while deciding the last phoneme of the incoming word. The vocabulary used for this work contains words having the last phoneme any of the consonant Assamese phonemes. Two class PNNs, which are trained to learn clean phonemes patterns, are again used to recognize the last phoneme from the SOM segmented phonemes of the incoming word. The codebook designed by LVQ has a distinct code for every word of the vocabulary. Suppose the PNN decides about a phoneme to be the last phoneme, then at the same time, the algorithm will search in the dictionary, whether any such word exists in the dictionary or not. But if no words of the vocabulary ends with that phoneme, then the decision is discarded. Thus, the codebook assistance assures that most likely decision about the last phoneme can be obtained.

7.5 Experimental Details and Results

The work is carried out as per the flow diagram of Fig. 7.2. Mathematically, the problem in hand can be stated as below

Table 7.1 PNN-based decision algorithm for vowel phoneme

1. Input: Speech S of size m× n, sampling frequency f_s, duration T second
2. Preprocess the signal using preprocessing algorithms described in Sect. 6.6.2
3. Obtain Dw1, Dw2, Dw3, Dw4, Dw5, Dw6 using the DWT based segmentation described in Sect. 6.4
4. Obtain SW1, SW2, SW3, SW4, SW5, SW6 using the SOM weight vector extraction algorithm described in Sect. 6.5.2
5. Find first formant frequency of Dw1, Dw2, Dw3, Dw4, Dw5, Dw6 and store as F_{DW1}, F_{DW2}, F_{DW3}, F_{DW4}, F_{DW5} and F_{DW6}
6. Find first formant frequency of SW1, SW2, SW3, SW4, SW5, SW6 and store as F_{SW1}, F_{SW2}, F_{SW3}, F_{SW4}, F_{SW5} and F_{SW6}

4. Load **PNN1**

6. Decide PHON-A
   If
   Dw1=PHON-A and F_{DW1}=F1 of phoneme /X/
   else if
   Dw2= PHON-A and F_{DW2}=F1 of phoneme /X/
   else if
   Dw3= PHON-A and F_{DW3}= F1 of phoneme /X/
   else if
   Dw4=PHON-A and F_{DW4}= F1 of phoneme /X/
   else if
   Dw5=PHON-A and F_{DW5}= F1 of phoneme /X/
   else if
   Dw6=PHON-A and F_{DW6}= F1 of phoneme /X/
   else Decide
   'Not Assamese Phoneme /X/'.
7. or Decide PHON-A
   If SW1=PHON-A and F_{SW1}=F1 of vowel /X/
   else if
   SW2= PHON-A and F_{SW2}= F1 of phoneme /X/
   else if
   SW3= PHON-A and F_{SW3}= F1 of phoneme /X/
   else if
   SW4=PHON-A and F_{SW4}= F1 of phoneme /X/
   else if
   SW5=PHON-A and F_{SW5}= F1 of phoneme /X/
   else if
   SW6=PHON-A and F_{SW6}= F1 of phoneme /X/
   else Decide
   'Not Assamese Phoneme /X/'.

```
Suppose,
```
C_i → Initial Consonant phoneme which can be vary within all the phoneme families
V_j → Vowel Phoneme
C_k → Last Consonant phoneme
Then any incoming word may have the form
$X = C_i V_j C_k$
where, i = 1 to 20 (excluding /η/, /w/, /j/)
j = 1 to 8
k = 1 to 23
Identify, C_i, V_j , and V_j.

The following section provides a brief account on each of the constituent blocks, with a description of the raw speech samples.

7.5.1 Experimental Speech Signals

The experimental speech samples are recorded in three phase. The clean consonant and vowel phonemes which are recorded from five girl speakers and five boy speakers are used to train PNNs. The description included in Chap. 4 simply reveals the fact that Assamese consonant phonemes can be classified into six distinct phoneme families as Unaspirated, Aspirated, Spirant, Nasal, Lateral, and Trill, excluding the semivowels /w/ and /j/. The sample words selected for this work cover variation in the initial phoneme from different families as well as variation of all the vowel phonemes. Therefore, for the second phase of recording, a word list is prepared as in Tables 7.2 and 7.3, which are recorded from the same girl and boy speaker that was used in the first phase. The third phase of sample collection covers some more samples recording with mood variations, for testing the ANNs. For recording the speech signal, a PC headset and a sound-recording software, Gold Wave, are used. The recorded speech sample has the following specifications:

- Duration = 2 s
- Sampling rate = 8,000 samples/second and
- Bit resolution = 16 bits/sample.

7.5.2 RNN Training Consideration

A RNN is constituted for this work with one hidden layer and tan-sigmoid activation functions. The input layer size is equal to the length of the feature vector, and the output layer is equal to the number of classes, which is the number of Assamese phoneme family, i.e., six. The F1 difference in initial phonemes divides the words

Table 7.2 CVC-type word list of for spoken word recognition model

S. No.	Vowel (V)	Initial consonant (C)	Last consonant (C)	CVC word
1	/i/	/x/	/t/	/xit/
2			/r/	/xir/
3			/k/	/xik/
4			/s/	/xis/
5	/u/	/d/	/kɦ/	/dukɦ/
6			/r/	/dur/
7			/t/	/dut/
8			/b/	/dub/
9	/e/	/b/	/x/	/bex/
10			/d/	/bed/
11			/s/	/bes/
12			/l/	/bel/
13	/o/	/m/	/n/	/mon/
14			/k/	/mok/
15			/r/	/mor/
16			/t/	/mot/

Table 7.3 CVC-type word list of for spoken word recognition model

S. No.	Vowel (V)	Initial consonant (C)	Last consonant (C)	CVC word
17	/ɛ/	/kɦ/	/l/	/kɦɛl/
18			/p/	/khɛp/
19			/d/	/kɦɛd/
20			/r/	/kɦɛr/
21	/ɑ/	/r/	/n/	/rɑn/
22			/kɦ/	/rAkɦ/
23			/tɦ/	/rAtɦ/
24			/x/	/rɑx/
25	/a/	/l/	/ bɦ/	/labɦ/
26			/z/	/laz/
27			/tɦ/	/latɦ/
28			/kɦ/	/lakɦ/
29	/ɔ/	/n/	/l/	/nɔl/
30			/d/	/nɔd/
31			/kɦ /	/nɔkɦ/
32			/m/	/nɔm/

among various phoneme family and that was used as the decision boundary among the six classes. The RNN training is carried out using (error) backpropagation with adaptive learning rate algorithm that updates weight and bias values according to gradient descent. The RNN training time is highly affected by the LPC order. The training time increases with the increase in LPC order. But lower order of LPC reduces the success rate during testing. RNN success rate for three different predictor size is shown in Table 7.4. According to [10], the RNN configuration with 40 hidden

Table 7.4 Performance of RNN versus predictor size

S. No.	RNN training time in sec	Predictor size	Success rate (%)
1	2545.012	90	85
2	1546.037	40	70
3	1380.034	10	60

Table 7.5 Phoneme family versus performance of RNN

S. No.	Phoneme family	Success rate (%)
1	*Unaspirated*	100
2	*Aspirated*	90
3	*Spirant*	90
4	*Nasal*	85
5	*Trill*	100
6	*Lateral*	100

Table 7.6 Overall success rate for vowel phoneme

S. No.	Vowel phoneme	Success rate (%)
1	*/i/*	98
2	*/e/*	96
3	/E/	94
4	*/a/*	98.5
5	/A/	92
6	/O/	91
7	*/o/*	97
8	*/u/*	95

neurons gave the best time and success rate combination. The RNN success rates for the various phoneme families are as shown in Table 7.5. The data shown in Table 7.5 are obtained for fixed predictor size of 40, which has been found to be a proper compromise between LPC predictor size and training time.

7.5.3 Phoneme Segmentation and Classification Results

The SOM is trained for six different iterations which provides identical number of decision boundaries. The segmented phonemes are then checked one by one to find which particular segment represents the vowel phoneme by matching pattern with the trained PNNs. The recognition rate for various vowels is shown in Table 7.6. Similarly, Table 7.7 shows success rate for various consonant phoneme as initial phoneme and last phoneme. Table 7.8 summarizes this performance difference for vowel, initial consonants, and last consonants.

Table 7.7 Overall success rate for consonant phoneme

S. No.	Phoneme	Success rate of initial phoneme (%)	Success rate of last phoneme (%)
1	/p/	×	85
2	/b/	90	100
3	/t/	×	90
4	/d/	97	90
5	/k/	×	85
6	/g/	×	×
7	/pH/	×	×
8	/bH/	×	89
9	/tH/	×	100
10	/dH/	×	×
11	/kH/	91	100
12	/gH/	×	
13	/s/	×	90
14	/z/	×	85
15	/x/	91	95
16	/H/	×	×
17	/m/	93	87
18	/n/	91	100
19	/l/	93	95
20	/r/	93	85

Table 7.8 Overall success rate for various phoneme

S. No.	Phoneme	Success rate (%)
1	Vowel	95.25
2	Initial consonant	94
3	Last consonant	93.2

7.6 Conclusion

This chapter explains a prototype model for recognition of CVC-type Assamese words, which can be later extended to develop a complete phoneme-based speech recognition model exclusively for Assamese language. The method uses a new SOM-based technique to segment the word into various phoneme boundaries which provides better recognition success rate than earlier reported results. The strength of the proposed algorithm lies in the fact that the uniqueness of Assamese phonemical structure is used effectively all throughout the work. Although computational time of the proposed method is more, it can be further reduced by improved training methods and distributed and parallel processing.

References

1. Eysenck MW (2004) Psychology-an international persepective. Psychology Press, Hove
2. Jusczyk PW, Luce PA (2002) Speech perception and spoken word recognition: past and present. Ear Hear 23(1):2–40
3. Lima Araujo AM de, Violaro F (1998) Formant frequency estimation using a Mel scale LPC algorithm. In: Proceedings of IEEE international telecommunications symposium, vol 1, pp 207–212
4. Snell RC, Milinazzo F (1993) Formant location from LPC analysis data. IEEE Trans Speech Audio Process 1(2):129–134
5. Rabiner LR, Schafer RW (2009) Digital processing of speech signals. Pearson Education, Dorling Kindersley (India) Pvt. Ltd, Delhi
6. Demuth H, Beale M, Hagan M, Venkatesh R (2009) Neural network toolbox6, users guide. Available via http://filer.case.edu/pjt9/b378s10/nnet.pdf
7. Bullinaria JA (2000) A learning vector quantization algorithm for probabilistic models. In: Proceedings of EUSIPCO, vol 2, pp 721–724
8. McClelland JL, Mirman D, Holt LL (2006) Are there interactive processes in speech perception? Trends Cogn Sci 10(8):363–369
9. Goswami GC (1982) Structure of Assamese, 1st edn. Department of Publication, Gauhati University, Assam
10. Sarma M, Dutta K, Sarma KK (2009) Assamese numeral corpus for speech recognition using cooperative ANN architecture. Int J Electr Electron Eng 3(8):456–465

Chapter 8
Application of Clustering Techniques to Generate a Priori Knowledge for Spoken Word Recognition

Abstract In this chapter, a technique is proposed to remove the CVC-type word limitation observed in case of spoken word recognition model described in Chap. 7. This technique is based on a phoneme count determination block based on K-means clustering (KMC) of speech data. Sections 8.2 and 8.3 of this chapter provides detail description of a K-mean algorithm-based technique to provide prior knowledge about the possible number of phonemes in a word. Experimental work, related to the proposed technique, is discussed in Sect. 8.4. Section 8.5 concludes the description.

Keywords K-mean · Clustering · Spoken word · RNN · Multiple phoneme

8.1 Introduction

The spoken word recognition model described in the Chap. 7 has the limitation that it can recognize only CVC-type words. But to develop a complete spoken word recognition model for the language of application which in this case is Assamese, the system should have the capability of dealing with words having different CV combination. Therefore, in order to further improve the model enabling it to deal with multiple phonemes, an extension to its present form should be given. The resulting framework incorporates the ability to provide certain a priori decision regarding the number of phonemes in any incoming word. This chapter proposes a solution to phoneme count with a priori knowledge using K-means clustering (KMC) and RNN. The problem is summarized in Fig. 8.1. The work described in the Chap. 7 has the noticeable limitation of recognizing only CVC-type words but formulates the basic principles required to perform ANN-based spoken word recognition. As depicted by Fig. 8.1, the basic requirement is to switch the algorithm from three-phoneme to four-phoneme words, four-phoneme to five-phoneme words and so on. Thus, the algorithm needs some prior knowledge about the number of phonemes in the word.

M. Sarma and K. K. Sarma, *Phoneme-Based Speech Segmentation Using Hybrid Soft Computing Framework*, Studies in Computational Intelligence 550,
DOI: 10.1007/978-81-322-1862-3_8, © Springer India 2014

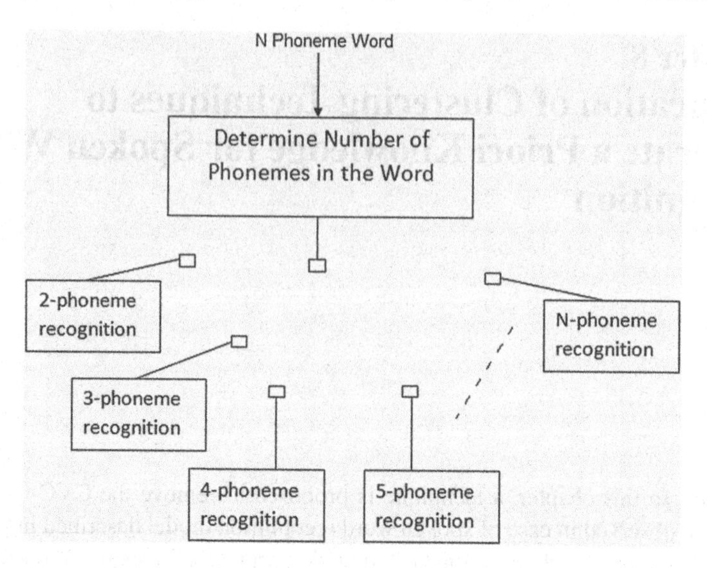

Fig. 8.1 Prospective scenario of the proposed spoken word recognition model

If we choose some two-phoneme, three-phoneme, four-phoneme, and five-phoneme sample word of varying CV combination and cluster them using KMC technique, where k-value will be the number of phonemes in each case, an RNN can be trained to learn the number of clusters in each case. Any word coming to the recognition algorithm will be first supported by a 2-mean clustering. The outcome shall be used by an RNN which provides class discrimination to classify it into 2-phoneme word. If RNN fails to classify it, then it is send for 3-mean clustering where the RNN attempts to classify the clustered data into 3-phoneme word and so on. If RNN fails to classify the word into any of the two-phoneme, three-phoneme, or four-phoneme combinations up to N-phoneme, then that word will be discarded. Otherwise, it will be send for respective N-phoneme word recognition process.

8.2 K-Means Clustering (KMC)

Clustering is one of the most important unsupervised learning problem. Literally, it can be defined as the process of organizing objects into groups where members are similar in some way. A cluster is, therefore, a collection of objects which have similar attributes between them and are dissimilar to the objects belonging to other clusters. In a more appropriate way, it can be said that the goal of clustering is to determine the intrinsic grouping in a set of unlabeled data. K-means is an exclusive clustering algorithm. The procedure involved in KMC follows a simple and easy way to classify a given data set through a certain number of clusters, suppose k. Value of k is fixed a priori. The main idea is to define k centroids, one for each cluster.

These centroids should be placed in an appropriate way because different location causes different results. So, the better choice is to place them as much as possible far away from each other. The next step is to take each point belonging to a given data set and associate it to the nearest centroid. When no point is pending, the first step is completed and an early grouping is done. At this point, the algorithm needs to recalculate k new centroids as barycenters of the clusters resulting from the previous step. After getting these k new centroids, a new binding has to be done between the same data set points and the nearest new centroid. A loop has been generated. As a result of this loop, the k centroids change their location step by step until no more changes are done. In other words, centroids do not move any more. Finally, this algorithm aims at minimizing an objective function, which in this case a squared error function. The objective function given by Eq. (8.1).

$$J = \sum i = 1^k \sum i = 1^n \parallel x_i^j - C_j \parallel^2 \tag{8.1}$$

where $\parallel x_i^j - C_j \parallel^2$ is a chosen distance measure between a data point x_i^j, and the cluster center C_j is an indicator of the distance of the n data points from their respective cluster centers [1, 2]. The algorithm can be summarized by the following steps:

1. Place K points into the space represented by the objects that are being clustered. These points represent initial group centroids.
2. Assign each object to the group that has the closest centroid.
3. When all objects have been assigned, recalculate the positions of the k centroids.
4. Repeat steps 2 and 3 until the centroids no longer move. This produces a separation of the objects into groups from which the metric to be minimized can be calculated.

The advantage of KMC is that with a large number of variables, k-means may be computationally faster, if k is small. K-means may produce tighter clusters than hierarchical clustering, especially if the clusters are globular.

8.3 KMC Applied to Speech Data

As already discussed in Sect. 8.2 clustering partitions or groups a given set of patterns into disjoint set or clusters. This is done such that patterns in the same clusters are alike, and patterns belonging to two different clusters are different. Clustering has been a widely studied problem in a variety of application domains including ANNs, artificial intelligence, and statistics. The number of clusters k is assumed to be fixed in KMC.

In this chapter, we propose a probable technique of generating some prior knowledge about number of phonemes in an word, without segmentation. The method involves clustering N-phoneme words into N-cluster using KMC. Here, N is the value of k in the KMC. Suppose that we have n speech frames $s_1, s_2, \cdots, s_n$ all

Table 8.1 KMC algorithm for speech data clustering

1. Input: Speech S of size $m \times n$, sampling frequency f_s, duration T second
2. Preprocess the signal using preprocessing algorithms described in Sect. 6.6.2
3. Set the number of clusters $K = N$;
4. Initialize cluster centroids as $\mu_1, \mu_2 \ldots \mu_k$ in the data space represented by the speech frames.
5. Assign each frame s_i to a group that has the closest centroid as follows

   Repeat until $C^{(i)}$ no longer move

   For every i, set

   $$C^{(i)} = argmin \parallel s^{(i)} - \mu_j \parallel^2$$

   Update centroid position $C^{(i)}$.

   Compute new centroid position from assigned frames.

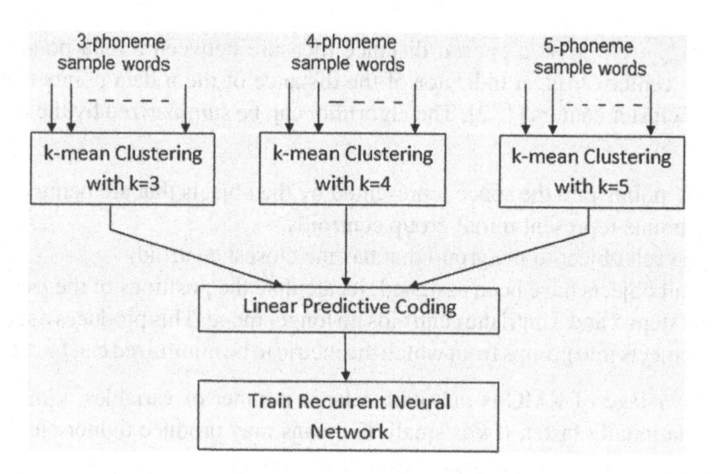

Fig. 8.2 Word clustering technique

from the same speech signal or one spoken word and we know that the word has N number of phonemes. Then, we will use KMC algorithm given in Table 8.1 to make $k = N$ number of clusters in the data set consisting of the speech frames $k < n$. Table 8.1 represents the KMC algorithm as it is used in this work. Let μ_i be the mean of the vectors in cluster i. If the clusters are well separated, then calculate minimum distance that the frames can be separated. That is, we can say that frame s_i is in cluster i if $| x - mi |$ is the minimum of all the k distances.

After getting the clustered data, LP coefficients are computed to reduce the size of the data vector and an RNN is trained to classify spoken words according to the number of clusters present. The word clustering technique using KMC algorithm proposed here can be visualized from the block diagram of Fig. 8.2.

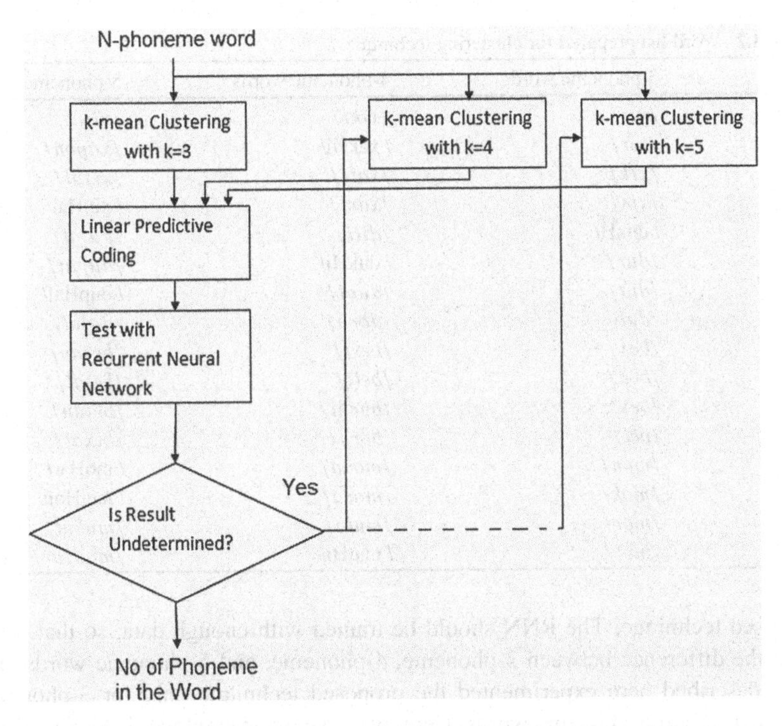

Fig. 8.3 Process logic of the proposed number of phoneme determination technique

8.4 Experimental Work

The experimental work is carried out as per the flow diagram of Fig. 8.3. Any word coming to the word recognizer is first clustered with KMC algorithm for k-value 3. Then, LP coefficients of the clustered data are extracted and presented to the RNN for classification. The RNN is trained to learn number of cluster in the data. If RNN cannot classify the data into any of the defined class, then the word is clustered with KMC algorithm for *k*-value 4 and the process is repeated. If the RNN again fails to classify the data, then the word is clustered with *k*-value 5 and the same process is repeated again. In this way, the proposed logic is used to determine the possible number of phonemes in the word for 3-phoneme, 4-phoneme, and 5-phoneme words. If the RNN fails to take any decision, then that particular word is discarded.

8.4.1 Experimental Speech Samples

Selection of proper set of sample words as a representative of all the N-phoneme words where (N = 2, 3 or 4) is an important factor for better success rates of the

Table 8.2 Word list prepared for clustering technique

S. No	3-phoneme words	4-phoneme words	5-phoneme words
1	/*xit*/	/*xiki*/	/*xital*/
2	/*xir*/	/ xakHi/	/*xapon*/
3	/*xik*/	/*xati*/	/*xijal*/
4	/*xis*/	/*xari*/	/ xipHal/
5	/ dukH/	/*dili*/	/*datal*/
6	/*dur*/	/ dukHi/	/*dupat*/
7	/*dut*/	/*duni*/	/ dupHal/
8	/*dub*/	/*dora*/	/*dojal*/
9	/*bex*/	/*besi*/	/*besan*/
10	/*bed*/	/*beli*/	/*betal*/
11	/*bes*/	/*burhi*/	/*betan*/
12	/*bel*/	/*bora*/	/*bixal*/
13	/*mon*/	/*mona*/	/ moHar/
14	/*mok*/	/*mora*/	/ metHon/
15	/*mor*/	/*mula*/	/*mukut*/
16	/*mot*/	/ mutHi/	/*mauzar*/

proposed technique. The RNN should be trained with enough data, so that it can learn the difference between 3-phoneme, 4-phoneme, and 5-phoneme words. The work described here experimented the proposed technique only for 3-phoneme, 4-phoneme, and 5-phoneme words. Here, we have considered words having all the Assamese vowel variations. Tables 8.2 and 8.3 show the word list. But to obtain better success rate, more words have to be included, such that all the possible vowel and consonant variation of Assamese vocabulary is covered. The words are recorded from two girl and boy speaker. Out of all collected data, 50 % of the words are used for training and 50 % are used for testing the clustering technique.

8.4.2 Role of RNN in Decision Making of the Proposed Technique

An RNN is playing the role of decision maker block by learning number of clusters in a sample data. The sample data of 3-phoneme words have 3 clusters, 4-phoneme words have 4 clusters, and 5-phoneme words have 5 clusters and so on. Accordingly, the RNN classifies the data into 3-cluster, 4-cluster, and 5-cluster etc. Here, we have considered only three variations of phonemes, 3 phonemes, 4 phonemes, and 5 phonemes to complete the design of the system and validate its abilities. Now, any N-phoneme word obtained from the recognition vocabulary shall be first clustered with $K = 3$, $K = 4$, and $K = 5$ and stored as DK3, DK4, and DK5, respectively. At first, DK3 is presented to the trained RNN to classify into any of the three classes. If RNN fails to classify, then DK4 and DK5 are presented consequently. If any of DK3, DK4, and DK5 do not come under any of the classes defined by RNN, then that word is discarded. The decision-making process can be described by the algorithm given in Table 8.4.

Table 8.3 Word list prepared for clustering technique

S. No	3-phoneme words	4-phoneme words	5-phoneme words
17	/kHEl/	/kHeda/	/kHAbAr/
18	/kHEp/	/bHada/	/kHANAl/
19	/kHEd/	/dHara/	/kHorak/
20	/kHEr/	/tHHka/	/kHoraN/
21	/rAn/	/roza/	/rAtAn/
22	/rAtH/	/ronga/	/rAbAr/
23	/rAkH/	/rati/	/rAgAr/
24	/rAx/	/roa/	/rAzAn/
25	/labH/	/lopHA/	/lAgAn/
26	/laz/	/lora/	/lAbAr/
27	/latH/	/lora/	/lasit/
28	/lakH/	/lAni/	/lAgAr/
29	/nOl/	/nAga/	/nAzAr/
30	/nOd/	/nila/	/nAtun/
31	/nOkH/	/nura/	/nAdAn/
32	/nOm/	/nija/	/nAgAr/

8.4.3 Result and Limitation

When the KMC-based phoneme count determination technique is introduced, the success rate of the word recognition system depends on the precision of clustering of speech data. Figure 8.4 represents a view of clustered data for a 3-phoneme word. The success of correct decision depends on several factors. Varying LP predictor length significantly effects the success rate and training time. Table 8.5 shows overall success rate for 3-phoneme, 4-phoneme, and 5-phoneme words with varying predictor length and corresponding RNN training time. As can be seen from Table 8.5, with 30 predictor size, we obtain better success rate. Similarly, with 30 predictor size, the RNN testing time for 3-phoneme, 4-phoneme, and 5-phoneme words are shown in Table 8.6. Table 8.7 shows success rate for 3-phoneme, 4-phoneme, and 5-phoneme words with fixed predictor size 30.

From the experimental results, it is clear that the proposed method can give a maximum of 83 % success rate only. Obviously, in order to use the logic in the proposed spoken word recognition method, the success rate must be improved, otherwise the whole recognition system shall suffer. Since speech signal provides some highly correlated data, KMC algorithm at times fails to capture that temporal variation. Therefore, the logic explained here requires further improvement, so that it can deal with high correlation of the speech data. After including the KMC algorithm to obtain a priori knowledge regarding the number of phonemes, the system overall training time and testing time increases. This is summarized in Table 8.8. This increase in the system execution time indicates that the KMC raises the computational complexity of the overall spoken word recognition block. Such limitations are to be taken care of as follow-up to the present work.

Table 8.4 KMC algorithm for speech data clustering

1. Input: Spoken word siignal S of size $m \times n$, sampling frequency f_s, duration T second
2. Preprocess the signal using preprocessing algorithms described in Sect. 6.6.2
3. Apply KMC algorithm described in Sect. 8.3 for $K = 3$;
4. Extract LP coefficients from the clustered data for P prediction order.

prediction order.

5. Present it to the trained RNN for classification.

    If
    Result is undetermined
    then
    Apply KMC algorithm described in Sect. 8.3 for $K = 4$
    Extract LP coefficients from the clustered data for P prediction order.
    Present it to the trained RNN for classification.
        If
        Result is undetermined
        then
        Apply KMC algorithm described in Sect. 8.3 for $K = 5$
        Extract LP coefficients from the clustered data for P prediction order.
        Present it to the trained RNN for classification.
            If
            Result is undetermined
            then
            Discard that word.
            else
            Output: N=Number of cluster. Send it to the N phoneme word recognition model
        else
        Output: N=Number of cluster. Send it to the N phoneme word recognition model
    else
    Output: N=Number of cluster. Send it to the N phoneme word recognition model

Table 8.5 Performance of RNN versus predictor size for clustering technique

S. No	RNN training time (s)	Predictor size	Success rate (%) Rate
1	2232.12	50	75.4
2	1546.037	30	81

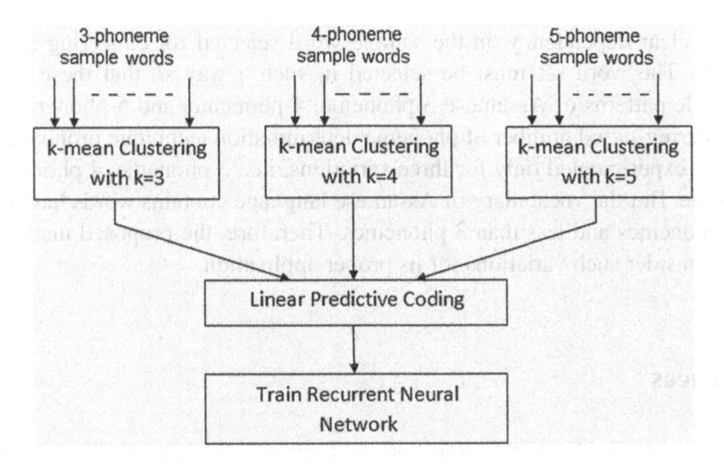

Fig. 8.4 Clustering using KMC for 3-phoneme

Table 8.6 RNN testing time with varying number of phonemes in the word

S. No	Word	RNN training time (s)
1	3-phoneme	1032.12
2	4-phoneme	1216.7
3	5-phoneme	1434.01

Table 8.7 Success rate of number of phoneme determination

S. No	Word	Success rate (%)
1	3-phoneme	83
2	4-phoneme	67.3
3	5-phoneme	78.1

Table 8.8 Execution time of the system with and without KMC

S. No		Overall training time (s)	Overall testing time (s)
1	System without KMC	1380.034	280.12
2	System with KMC	4380.02	1532.4

8.5 Conclusion

In this chapter, a method is described in order to take some prior decision about the probable number of phonemes in a spoken word, so that the spoken word recognition model can be further improved to include multiple-phoneme words. But the results obtained from the proposed KMC algorithm-based technique need improvement, in order to obtain better success rates. Further, the success rate of the proposed logic has clear dependency on the sample word selected for clustering and training RNN. The word set must be selected in such a way so that these cover all the syllable patterns of Assamese 3-phoneme, 4-phoneme, and 5-phoneme words. The clustering-based number of phoneme determination technique proposed in this chapter is experimented only for three variations, i.e., 3 phoneme, 4 phoneme, and 5 phoneme. But the vocabulary of Assamese language contains words having more than 5 phonemes and less than 3 phonemes. Therefore, the proposed method must have to consider such variations for its proper application.

References

1. Tapas Kanungo T, Mount DM, Netanyahu NS, Piatko CD, Silverman R, Wu AY (2002) An efficient k-means clustering algorithm: analysis and implementation. IEEE Trans Pattern Anal Mach Intell 24(7):881–892
2. Wagstaff K, Cardie C, Rogers S, Schroedl S (2001) Constrained K-means clustering with background knowledge. In: Proceedings of the 18th international conference on machine learning, pp 577–584

Chapter 9
Application of Proposed Phoneme Segmentation Technique for Speaker Identification

Abstract This chapter presents a neural model for speaker identification using speaker-specific information extracted from vowel sounds. The vowel sound is segmented out from words spoken by the speaker to be identified. Vowel sounds occur in a speech more frequently and with higher energy. Therefore, situations where acoustic information is noise corrupted, vowel sounds can be used to extract different amounts of speaker discriminative information. The model explained here uses a neural framework formed with PNN and LVQ where the proposed SOM-based vowel segmentation technique is used. The work extracts glottal source information of the speakers initially using LP residual. Later, empirical-mode decomposition (EMD) of the speech signal is performed to extract the residual. Depending on these residual features a LVQ-based speaker code book is formed. The work shows the use of residual signal obtained from EMD of speech as a speaker discriminative feature. The neural approach of speaker identification gives superior performance in comparison with the conventional statistical approach like hidden Markov models (HMMs), Gaussian mixture models (GMMs), etc. found in the literature. Although the proposed model has been experimented in case of the speakers of Assamese language, it shall also be suitable for other Indian languages for which the speaker database should contain samples of that specific language.

Keywords Speaker · Identification · ANN · Codebook

9.1 Introduction

Phonemes are linguistic abstraction with a high degree of variation of length in the acoustic speech signal and therefore are difficult to differentiate into distinct segments. Acoustic appearance of phonemes varies according to their context as well as from speaker to speaker. Vowel phonemes are a part of any acoustic speech signal. Vowel sounds occur in a speech more frequently and with higher energy. Therefore, vowel phoneme can be used to extract different amounts of speaker discriminative information in situations where acoustic information is noise corrupted. Use of vowel

M. Sarma and K. K. Sarma, *Phoneme-Based Speech Segmentation Using Hybrid Soft Computing Framework*, Studies in Computational Intelligence 550,
DOI: 10.1007/978-81-322-1862-3_9, © Springer India 2014

sound as a basis for speaker identification has been initiated long back by the Speech Processing Group, University of Auckland, New Zealand [1]. Since then, phoneme recognition algorithms and related techniques have received considerable attention in the problem of speaker recognition and have even been extended to the linguistic domain. Role of vowel phoneme is yet an open issue in the field of speaker verification or identification. This is because of the fact that vowel phoneme-based pronunciation vary with regional and linguistic diversity. Hence, segmented vowel speech slices can be used to track regional variation in the way the speaker speaks the language. It is more so in case of a language like Assamese spoken by over three million people in the northeast state of Assam with huge linguistic and cultural diversity which have influenced the way people speak the language. Therefore, an efficient vowel segmentation technique shall be effective in speaker identification as well as for applications like speech to text conversion and voice-activated system.

ANN model has recently come into focus in the field of speaker recognition, due to its potential to demonstrate excellent results in certain frameworks where machine learning and higher computational requirements dictate outcomes in a parallel and high computational approach [2–9]. The brain's impressive cognitive skills have motivated the researchers to explore the possibility of ANN models configured specially for the field of speech and speaker recognition in 1980s [10, 11] and has been driven by a hope that bioinspired neural network-like models may ultimately lead to human-like performance while dealing with such complex tasks. But at the later half of 1990s, suddenly, ANN-based speech research lost the attraction and was nearly terminated [10]. Statistical frameworks like hidden Markov models (HMMs), Gaussian mixture models (GMMs), etc. come into focus [12–16], supporting both acoustic and temporal modeling of speech. However, it should be mentioned that the best current systems are far from equaling human-like performance and many important research issues are still to be explored. Therefore, the ANN-based research is receiving renewed attention and is now considered as a sought-after area in emerging issues in speech and speaker recognition.

The purpose of a speaker identification system is to determine the identity of an unknown speaker among several with known speech characteristics. The identity can be established from a sample of his or her voice. Speaker identification is divided into two categories: closed set and open set. A closed-set speaker identification system identifies the speaker as one of those enrolled, even if he or she is not actually enrolled in the system. On the other hand, an open-set speaker identification system should be able to determine whether a speaker is enrolled or not (impostor) and, if enrolled, determine his or her identity. The task can also be divided into text-dependent and text-independent identification. The difference is that in the first case, the system knows the text spoken by the person, while in the second case, the system must be able to recognize the speaker from any text [17]. This chapter presents vowel sound-based closed-set text-independent speaker identification technique using the phoneme segmentation method explained in Chap. 6. The proposed SOM- and PNN-based phoneme segmentation technique is used to separate out the vowel phoneme from words spoken by some trained Assamese male and female speakers. A composite ANN framework provides different segments of the underlying structures and takes the most probable decision with the help of prior knowledge of the vowel pattern and characteristics. The speaker identification system is described in two parts, first using LP residual as speaker discriminative feature. The LP error sequence provides the speaker source information by subtracting the vocal tract effect, and therefore, it can be used as an effective feature for speaker recognition. In the next part, empirical-mode decomposition (EMD) algorithm is applied to obtain the residual. The EMD, first introduced by Huang et al. in 1998 [18], adaptively decomposes a signal into oscillating components and a residue. This residual is used as the speaker discriminative feature. One speaker code book is designed using these features. The LVQ code book contains some unique code for all the speakers in terms of vowel sound's source pattern. Speaker identification is carried out by first segmenting the vowel sounds from the speaker's speech signal and then matching the vowel source pattern with the LVQ code book. The proposed technique is novel due to the fact that such a composite ANN framework provides different segments of the underlying structures and takes the most probable decision with the help of prior knowledge of the vowel pattern and source characteristics. Very few works using such an ANN-based frameworks have been explored previously [4, 6–9]. A few works of similar nature are going on in different Indian languages [2, 3, 5], but to the best of our knowledge, no such speaker identification work has been reported in Assamese and other Indian languages so far, based on a vowel phoneme segmentation technique where the individual speaker's source characteristics and pattern of the vowel is used to differentiate speakers. Assamese is a widely spoken language in northeast India with vast linguistic diversity across different regions of the state and provides a sound area for research in phoneme-based speaker recognition. The objective of this work is to generate a platform for designing speaker recognition model taking advantage of such regional sound variations between speakers. The speaker database used in the work is developed from a set of speakers covering four different dialects of Assamese language, with gender and mood variations.

9.2 Certain Previous Work Done

In this section, a review of current speaker recognition work is provided. Initially, a few papers are described on speaker identification or verification focusing on a generalized scenario. Later, a few works are described which are specific to vowel- or voiced sound-based speaker recognition. Although the basic purpose of speaker identification and verification is different, but the discriminative features used are somewhat similar. Hence, this review covers both verification and identification systems.

1. In 1990, Templeton and Gullemin reported a work [11] on speaker identification based on vowel sounds using ANNs where the authors presented the results of an experiment. It applies MLPs to classify speakers using unweighted cepstral coefficients. They compared the results with [1] and found that the performance of MLPs is somewhat superior .

2. Fakotakis et al. in 1991 [19] reported a work on automatic text-independent speaker recognition based on spotting the stable part of the vowel phonemes of the test utterances, extracting parameter vectors and classifying them to a speaker-dependent vowel reference database. The system developed is suitable for identification as well as for verification purposes and was tested over a period of 4 months with a population of 12 male and female speakers with non-correlated training and test data. The accuracy of the system as measured by experimentation is satisfactory considering that the training utterances per speaker do not exceed 50 s and the test utterances 1 s in average.
3. In 1995, Thcvenaz and Hiigli reported a work [20] on speaker recognition using LP residue and analyzed its corresponding recognition performance by issuing experiments in the context of text-independent speaker verification. The experimental results show the usefulness of residue features as a speaker discriminative feature, and if synthesis filter features are combined with residue, then 5.7–4.0 % error reduction is achieved.
4. Another work is reported by Radova and Psutka in 1997 [21] where multiple classifiers are used for speaker identification. In this work, the attributes representing the voice of a particular speaker are obtained from very short segments of the speech waveform corresponding only to one pitch period of vowels. The patterns formed from the samples of a pitch period waveform are either matched in the time domain by use of a nonlinear time warping method, known as dynamic time warping (DTW), or they are converted into cepstral coefficients and compared using the cepstral distance measure.
5. Another work is reported by Sarma and Zue in 1997 [22], on a competitive segment-based speaker verification system using SUMMIT. They have modified SUMMIT further. The speech signal was first transformed into a hierarchical segment network using frame-based measurements. Acoustic models for 168 speakers were developed for a set of 6 broad phoneme classes which represented feature statistics with diagonal Gaussians, preceded by principle component analysis. The feature vector included segment-averaged MFCCs, plus three prosodic measurements: energy, fundamental frequency (F0), and duration. The size and content of the feature vector were determined through a greedy algorithm while optimizing overall speaker verification performance.
6. In 2003, Hsieh et al. reported another work on speaker identification system based on wavelet transform and GMM. The work presented an effective and robust method for extracting features for speech processing. Based on the time-frequency multiresolution property of wavelet transform, the input speech signal is decomposed into various frequency channels. For capturing the characteristics of the vocal tract and vocal cords, the traditional linear predictive cepstral coefficients (LPCC) of the approximation channel and the entropy of the detail channel for each decomposition process are calculated. In addition, a hard thresholding technique for each lower resolution is applied to remove interference from noise. The proposed feature extraction algorithm is evaluated on the MAT telephone speech database for text-independent speaker identification using the Gaussian mixture model (GMM) identifier [14].

7. In 2006, Prasanna et al. described a work [23] on speaker recognition using LP residual as speaker-specific information extracted from the excitation in the voiced speech. They have used AANN models to capture this excitation component of speech. They have demonstrated that for a speech signal sampled at 8 kHz, the LP residual extracted using LP order in the range 820 best represents the speaker-specific excitation information and the speaker recognition system using excitation information and AANN models requires significantly less amount of data both during training as well as testing, compared to the speaker recognition system using vocal tract information.
8. Espy-Wilson et al. in 2006 [24] reported a speaker identification system using speaker-specific information, a small set of low-level acoustic parameters that capture information about the speakers source, vocal tract size, and vocal tract shape. They demonstrated that the set of eight acoustic parameters has comparable performance to the standard sets of 26 or 39 MFCCs for the speaker identification task.
9. Antal has described another work on phonetic speaker recognition where the discriminative powers of broad phonetic classes for the task of speaker identification is discussed, and he has claimed that phonetic speaker models are more suitable for speaker recognition than standard models [25]. The work was reported in 2008.
10. Another work [26] on speaker identification in the SCOTUS corpus is reported by Jiahong and Mark in 2008, which includes oral arguments from the Supreme Court of the United States. They have reported that combination of GMM and monophone HMM models attains near-100 % text-independent identification accuracy on utterances that are longer than 1 s and the sampling rate of 11,025 Hz achieves the best performance. The work also shows that a sampling rate as low as 2,000 Hz achieves more than 90 % accuracy. Distance score based on likelihood numbers was used to measure the variability of phones among speakers. They have found that the most variable phone UH (as in good) and the velar nasal NG is more variable than the other two nasal sounds *M* and *N*. Their models achieved perfect forced alignment on very long speech segments.
11. Ferras et al. proposed a work [27] in 2009 where maximum likelihood linear regression (MLLR) transform coefficients have shown to be useful features for text-independent speaker recognition systems. In this work, they have used lattice-based MLLR. Using word lattices instead of 1-best hypotheses, more hypotheses is considered for MLLR estimation, and thus, better models are more likely to be used. As opposed to standard MLLR, language model probabilities are taken into account in this work. They have also shown that systems using lattice MLLR outperform standard MLLR systems in the Speaker Recognition Evaluation (SRE) 2006.
12. In 2010, Tzagkarakis and Mouchtaris reported two methods based on the generalized Gaussian density (GGD) and sparse representation classification (SRC) for noise-robust text-independent speaker identification and compared against a baseline method for speaker identification based on the GMM. They evaluated performance of each method in a database containing twelve speakers.

The main contribution of the work was to investigate whether the SRC and GGD approaches can achieve robust speaker identification performance under noisy conditions using short-duration testing and training data in relevance to the baseline method and indicated that the SRC approach significantly outperforms the other two methods under the short test and training sessions restriction, for all the signal-to-noise ratio (SNR) cases that were examined [28].

13. In 2011, another work [29] reported by Shimada et al. used pseudo pitch-synchronized phase information in voiced sound for speaker identification. Speaker identification experiments were performed using the NTT clean database and JNAS database. Using the new phase extraction method, they have obtained a relative reduction in the speaker error rate of approximately 27 and 46 %, respectively, for the two databases. They have obtained a relative error reduction of approximately 52 and 42 %, respectively, when combining phase information with the MFCC-based method.
14. Pradhan and Prasanna described the significance of information about vowel onset points (VOPs) for speaker verification in 2011 [30]. Vowel-like regions can be identified using VOPs. Vowel-like regions have impulse-like excitation, and therefore, impulse response of vocal tract system is better manifested in them and are relatively high-signal-to-noise-ratio (SNR) regions. They demonstrated that speaker information extracted from such regions is more discriminative.
15. Kinnunen et al. in 2011 reported a work [31] on speaker identification where feature vectors are extracted from the samples by short-term spectral analysis and processed further by vector quantization for locating the clusters in the feature space. A given speaker is matched to the set of known speakers in a database. The database is constructed from the speech samples of each known speaker. They have compared the performance of different clustering algorithms and the influence of the codebook size in this work.
16. Vuppala and Rao in 2012 explored the features extracted from steady vowel segments for improving the performance of speaker identification system under background noise [32]. In this work, steady vowel regions are determined by using the knowledge of accurate VOP and epochs. GMM-based modeling is explored for developing speaker models.

From this review, it is observed that following the work of Miles and Templeton and Gullemin in the University of Auckland, New Zealand, during 1990s, several researches focused on certain where acoustic parameters that capture information about the speakers source, vocal tract size, and vocal tract shape, etc. These are used as speaker-specific features. Few works use the vowel or voiced sounds to extract speaker discriminative feature. Fakotakis et al., Alafaouri et al., and Vuppala et al. use vowel spotting, formants of vowels, and steady vowel segments for speaker discrimination, respectively, whereas Pradhan et al. demonstrated the importance of VOPs recently. Shimada et al. used pseudo pitch-synchronized phase information in voiced sound for speaker identification. The trend is to capture more speaker-specific features enabling near-human-like recognition performance.

9.3 Linear Prediction Residual Feature

The LP error sequence or the LP residual, given by Eq. (7.5), provides the source information suppressing the vocal tract information obtained in the form of LP coefficients [33]. The source information obtained from the error sequence can be used as a relevant feature for speaker identification. The representation of source information in the LP residual depends upon the order of prediction. According to [33], for a speech signal sampled at 8 kHz, the LP residual extracted using the LP order in the range 8–20 best represents the speaker-specific source information. This work uses prediction order 20 to extract speaker-specific feature for both clean vowel and word, which have provided satisfactory success rates.

9.4 EMD Residual-Based Source Extraction

EMD is an adaptive method to analyze non-stationary signals which can be used to decompose nonlinear and non-stationary time series into a set of high-frequency mode called intrinsic mode functions (IMFs) and low-frequency component called the residual. The starting point of EMD is to locally estimate a signal as a sum of a local trend, i.e., the low-frequency part and a local detail, i.e., the high-frequency part. When this is done for all the oscillations composing a signal, the procedure is then applied again to the residual, considered as a new times series, extracting a new IMF and a new residual. The process continues until IMF satisfies the two IMF criteria. The first is that the number of extrema (sum of maxima and minima) and the number of zero crossing must be equal or differ by one. The second is the mean of the cubic splines must be equal to zero at all points [18, 34]. At the end of the decomposition process, the EMD method provides a signal as the sum of a finite number of IMFs and a final residual. The EMD algorithm for a given signal x(t) can be summarized as follows [18]:

1. Identify all extrema i.e., maxima and minima of $x(t)$;
2. Generation of the upper and lower envelope via cubic spline interpolation among all the maxima (resp. $emax(t)$) and minima (resp. $emin(t)$) respectively;
3. Compute the local mean series $m(t) = (emin(t) + emax(t))/2$;
4. Extract the detail $\mathrm{d}(t) = x(t) - m(t)$, where $\mathrm{d}(t)$ is the IMF candidate;
5. Iterate on the residual $m(t)$;

Practically, the above procedure is refined by a sifting process [18] which amounts to first iterating steps 1–4 upon the detail signal $d(t)$, until this latter can be considered as zero mean according to the stopping criterion. Once this is achieved, the detail is referred to as an IMF, the corresponding residual is computed, and step 5 applied.

Speech is the naturally obtained nonlinear and non-stationary form of signal. Therefore, linear analysis of speech signal has some potential risks like discarding nonlinear contents which shall lead to the loss of vital elements in the speech sample.

The end result shall be a mechanism which will have lower capability of human-like processing. Bioinspired computation possesses nonlinear attributes which contain a host of features enabling the brain to make best of decision with minutest of contents. Hence, linear processing of speech shall be detrimental to optimal decision making of a speech-based application. Therefore, linear analysis is replaced by the nonlinear processing observed in EMD. After introduction of EMD method by Huang et al. in 1998 [18], the speech researchers over the world carried out experiments to apply EMD process in speech analysis [35–39]. In this work, EMD is applied to decompose the vowel part of speech signal into intrinsic oscillatory modes and to obtain the residual signal. Since the residue is the lowest frequency component left after decomposition, it provides the frequency representation of the glottal source. We have extracted principal components from it to reduce the dimension, and these are used to train the LVQ network.

9.5 LVQ Codebook and Speaker Identification

A LVQ-based speaker code book is designed for the work using the principal components of the speaker-specific glottal source information in the form of LP or EMD residual. Manually marked vowels recorded from every speaker is directly used to extract the residual feature from LP model and EMD method. LVQ consists of two layers. The first-layer maps input vectors into clusters that are found by the network during training. The second-layer maps merge groups of first-layer clusters into the classes defined by the target data. Here, competitive layer learns to classify input vectors into target classes chosen by the user unlike strictly competitive layer possessed by SOM. Therefore, LVQ can be used to create code book. It models the discrimination function defined by the set of labeled code book vectors and the nearest neighborhood search between the code book and data. The code book training involves the LVQ1 algorithm described in Bullinaria [40]. The LVQ code book provides a unique code for every speaker based on the residual features.

9.6 Speaker Database

The speaker database is created from speakers of all the four different dialects of Assamese language. A word list is prepared containing 25 consonant–vowel–consonant (CiVCi and CiVCj) and 20 vowel–consonant–vowel (VCV and VCV) syllables with all the 8 vowel variations. These words are recorded by 5 speakers from each dialect, i.e., ($5 \times 4 = 20$) trained speakers in a room environment. Each CVC and VCV token is repeated five times, yielding a total of $45 \times 5 = 225$ tokens. The total 225 tokens are divided for two parts for PNN training and testing and for LVQ codebook. Here, 50 CVC and 50 VCV tokens are manually marked to get the vowel segments which are used for PNN training and testing, i.e., out of

(50 + 100 = 150) tokens, PNN is trained with 80 vowel segments and 70 vowel segments are used to test and fix the smoothing parameter. The rest 125 tokens are used for building the speaker code book. A separate database of 20 CVC-type words recorded in a somewhat noisy environment is used to test the speaker identification model. In both the cases, the speakers are same. For recording the speech signal, a PC headset and a speech analysis software, Wavesurfer, is used. The recorded speech sample has the following specification:

- Sampling rate = 16,000 samples/s,
- Bit resolution = 16 bits/sample.

9.7 Experimental Details and Results

The occurrence of vowel phonemes in spoken words of Assamese language varies with the change of dialects. Assamese native speakers use different vowel sound for the same word, when the speakers are from different region of the state. Use of vowel sound as a basis for speaker identification is a sound area of research in the field of speaker recognition. Therefore, this regional variation of occurrence of vowel sound in the same Assamese spoken word can be used as speaker discriminative feature for developing a Assamese speaker identification model. Such a prototype model is described in this work; Assamese speakers are identified using the vowel sound segmented out from words spoken by a speaker.

9.8 System Description

The proposed system can be stated by the following three points:

- The speaker to be identified is asked to utter a CVC word from which vowel part is extracted using the SOM- and PNN-based segmentation and vowel recognition technique.
- The extracted vowel is presented to the feature extraction block which provides speaker-specific glottal source information.
- Then, the principal components of the glottal source information is applied to the LVQ network for a pattern matching operation to identify the speaker. The LVQ network provides a code book of 20 native Assamese speakers, trained with their glottal source information obtained from manually marked vowel sounds.

The process logic of the proposed system is shown in Figs. 9.1 and 9.2 for LP residual and EMD residual, respectively. The following section describes the process in detail.

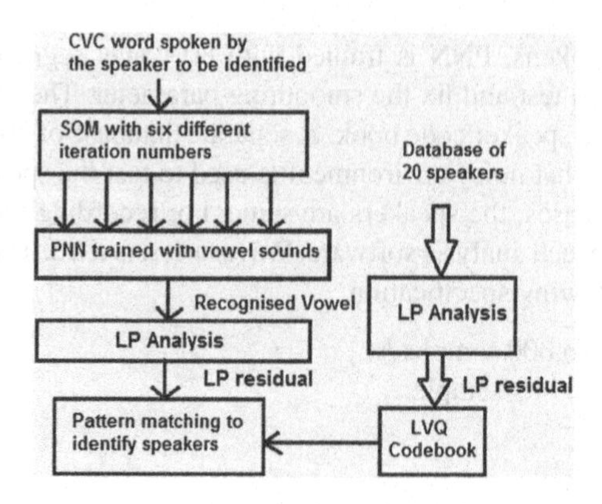

Fig. 9.1 Process logic of the speaker identification model using LP residual

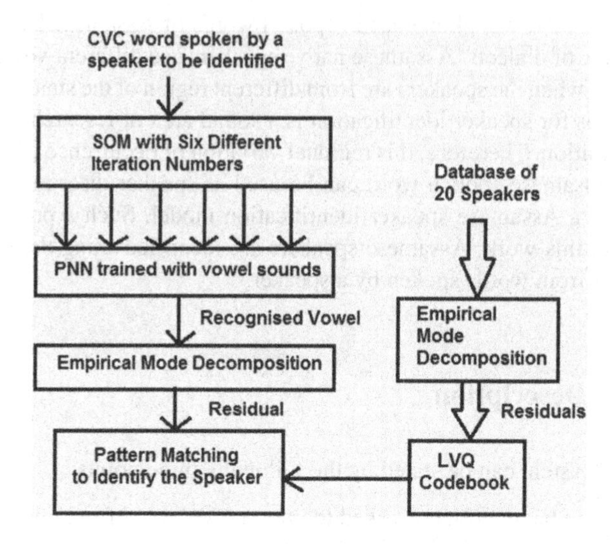

Fig. 9.2 Process logic of the speaker identification model using EMD residual

9.8.1 Vowel Segmentation Results

The vowel phoneme from any word spoken by the speaker to be identified is segmented and recognized by the SOM-based segmentation and PNN-based vowel recognition algorithm as described in Chap. 6. The description of the said segmentation algorithm is also included in Chap. 6, where LP coefficients of the speech samples are presented to the segmentation block. While applying to the speaker identification model, we have modified the algorithm a bit. This time, the speech signal is directly presented to the segmentation block. A brief description of the algorithm is included below to maintain the flow of the description.

SOM can be considered as a data visualization technique, i.e., it provides some underlying structure of the data. This idea is used in our vowel segmentation technique. If we train the same SOM with different epochs or iteration numbers, then SOM provides different weight vectors consisting of the wining neuron along with the neighbors. Thus, with different epochs, different internal segments or patterns of the speech signal can be obtained. Suppose, we have a one-dimensional field of neurons and say the samples of spoken word uttered by the speaker has a form

$$X_k = \left(x_1^k, x_2^k, \ldots, x_n^k\right) \tag{9.1}$$

When such an input vector is presented to the field of neurons, the algorithm will start to search the best matching weight vector W_i and thereby identify a neighborhood $\aleph_i$ around the wining neuron. While doing so, it will try to minimize the Euclidian distance $|X_k - W_i(k)|$. Adaptation of the algorithm will take place according to the relation,

$$W_i(k+1) = \begin{cases} W_i(k) + \eta_k(X_k - W_i(k)), & \text{i} \in \aleph_i^k; \\ W_i(k), & \text{i not } \aleph_i^k; \end{cases} \tag{9.2}$$

where learning rate η_k is having the form of

$$\eta_k = \eta_0 \left[1 - \frac{\text{epoch number}}{2 \times (\text{Size of the speech sample})}\right] \tag{9.3}$$

Thus, with the change of epoch number, different W_i will be obtained. Therefore, by training the same SOM block for various numbers of iterations, we get different weight vectors, each of which is considered as a segment of different phonemes constituting the word. From these segments, the relevant vowel portions are recognized by pattern matching done with some PNNs which are trained to learn the patterns of all Assamese vowel phonemes using some manually marked vowel part at an earlier time. PNNs handle data that have spikes and points outside the norm better than other ANNs. Therefore, PNN is suitable for problems like phoneme classification. Here, a two-class PNN-based classification is performed, where four PNNs are trained with two manually marked vowel phonemes and are named as *PNN1*, *PNN2*, *PNN3*, and *PNN4*, i.e., the output classes of *PNN1* are /i/ and /u/, the output classes of *PNN2* are /e/ and /o/, etc. These four PNNs are used sequentially to identify the segmented vowel phonemes. Manually marked vowels are used subsequently as the inputs in the input layer to the PNN enabling it to provide to each neuron the scope of learning the patterns and group them as per the derived decision. This happens in the pattern layer of the PNN.

Initially, four PNNs are trained with vowels segmented from CVC- and VCV-type Assamese words by manual marking in the speech analysis software PRAAT [41]. As mentioned in Sect. 9.5, out of (50 + 100 = 150) vowel segments obtained from manual marking, PNN is trained with 80 vowel segments and rest 70 vowel segments are used to test the PNN. Before using the PNN in the main speaker identification algorithm, it is tested with manually marked vowel segments. The smoothing parameter σ plays a very important role in proper classification of the input phoneme patterns during training. Since it controls the scale factor of the exponential activation function, its value should be the same for every pattern unit. A small value of σ causes the estimated parent density function to have distinct modes corresponding to the locations of the training samples. A larger value of σ produces a greater degree of interpolation between points. A very large value of σ would cause the estimated density to be Gaussian regardless of the true underlying distribution. However, it has been found that in practical problems, it is not difficult to find a good value of σ, and the misclassification rate does not change dramatically with small changes in σ. Table 9.1 shows the rate of correct classification for a range of σ value, and it can be seen that from 0.100, if we start to decrease, PNN shows 100 % correct classification, and therefore, PNN trained with these values are found to be useful for testing. For testing, a decision tree is designed using the four PNNs, where all the PNNs are checked one by one. The recognition rate of the four PNNs with $\sigma = 0.111$ while tested with manually marked vowel is shown in Table 9.2. The experiments are repeated for several trials and the success rates calculated.

After obtaining a acceptable success rate using the manually marked vowels, the same PNN decision tree is used to recognize SOM-segmented vowels. The vowel segmentation and recognition success rate using the SOM and PNN combination is summarized in Table 9.3.

As can be seen from Table 9.3, the success rate obtained from SOM-based segmentation is satisfactory. The success rate of the segmentation stage is measured using the performance of the PNN. Since the outcome of the PNN stage is validated using pattern match with manually extracted samples where above 90 % similarity is considered as a threshold. Therefore, SOM-provided vowel segments are proper for our purpose. We have fixed the SOM- and PNN-based vowel recognition block to use it in our speaker identification model.

9.8.2 Speaker Identification Results Using LP Residual

The segmented vowels are next applied for pattern matching with the trained LVQ code book which discriminates between speakers. Although, the subjects used for this work are twenty (ten male and ten female), it is used for testing eight Assamese vowels each segmented from five different words recorded under three to five different background noises as described in Sect. 9.6. The speaker identification success rate can be summarized by the Table 9.4 for two different predictor sizes. It can be seen from the experimental results that with 20 predictor size, the success rate shows around 3 % improvement in comparison with previous reported result [42]. Tables 9.5 and 9.6 show the correct speaker identification of segmented vowel from five different words containing the vowel in case of boy speaker-I and girl speaker-I, respectively. The correct identification of speakers directly depends on the segmentation process. If error occurs in the segmentation part, then obviously the speaker identification goes wrong.

Table 9.1 PNN classification rate for various σ value

S. No	σ	Correct classification (%)	Misclassification (%)
1	0.0001	100	0
2	0.001	100	0
3	0.01	100	0
4	0.1	100	0
5	0.111	100	0
6	0.222	100	0
7	0.333	93.75	6.25
8	0.444	93.75	6.25
9	0.555	93.75	6.25
10	0.666	87.5	12.25
11	0.777	81.25	18.75
12	0.888	81.25	18.75
13	0.888	81.25	18.75
14	0.999	81.25	18.75
15	1	81.25	18.75
16	2	75	25
16	2	75	25
17	10	75	25
18	30	75	25
16	50	75	25
16	100	75	25

Table 9.2 Success rate of PNN while tested with manually segmented vowels

S. No	Vowel	Correct recognition (%)	False acceptance (%)	Unidentified (%)
1	/i/	93.4	4	2.6
2	/e/	98.2	1.8	0
3	/ɛ/	94	2	3
4	/a/	97.5	2	0.5
5	/ɑ/	99	1	0
6	/ɔ/	98	2	0
7	/o/	92	6	2
8	/u/	96	4	0

Table 9.3 SOM segmentation success rate

S. No.	Vowel	Success rate of SOM (%)
1	/i/	98
2	/e/	96
3	/ɛ/	94
4	/a/	99
5	/ɑ/	99
6	/ɔ/	98
7	/o/	97
8	/u/	95

Table 9.4 Speaker identification success rate over all the vowel and word variations using LP residual

S. No.	Speaker	Success rate (Predictor size = 15) (%)	Success rate (Predictor size = 20) (%)
1	Boy speaker 1	88	95.2
2	Boy speaker 2	83.2	92.3
3	Boy speaker 3	87	99
4	Boy speaker 4	90	95
5	Boy speaker 5	80.6	95.1
6	Girl speaker 1	85	98
7	Girl speaker 2	87.8	93.7
8	Girl speaker 3	89	91.9
9	Girl speaker 4	91	94
10	Girl speaker 5	79.4	97.2
Total	Average	86.1	95.14

Table 9.5 Correct identification of boy speaker-I for LP residual

S. No.	Segmented vowel	Word-I	Word-II	Word-III	Word-IV	Word-V
1	/i/	√	√	√	√	×
2	/e/	×	√	√	√	√
3	/ɛ/	√	√	√	√	√
4	/a/	√	×	×	√	√
5	/ɑ/	×	√	√	√	√
6	/ɔ/	√	√	√	×	√
7	/o/	√	√	√	√	×
8	/u/	√	√	√	√	√

Table 9.6 Correct identification of girl speaker-I for LP residual

S. No.	Segmented vowel	Word-I	Word-II	Word-III	Word-IV	Word-V
1	/i/	×	√	√	√	×
2	/e/	√	√	√	√	×
3	/ɛ/	√	√	√	√	√
4	/a/	√	×	√	√	√
5	/ɑ/	×	√	√	√	√
6	/ɔ/	√	√	√	×	√
7	/o/	√	√	√	√	×
8	/u/	√	√	√	×	√

9.8.3 Speaker Identification Results Using EMD Residual

As mentioned earlier, EMD residual signal is used as speaker discriminative feature in the second part of the proposed speaker identification model. This time LVQ codebook is designed using the EMD residual signal obtained from 20 speakers.

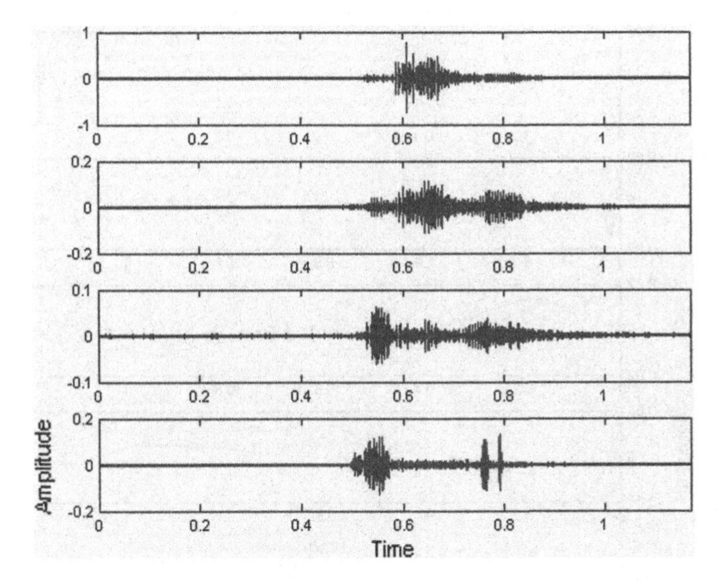

Fig. 9.3 IMF1 to IMF4 for vowel segment /e/ uttered by a male speaker

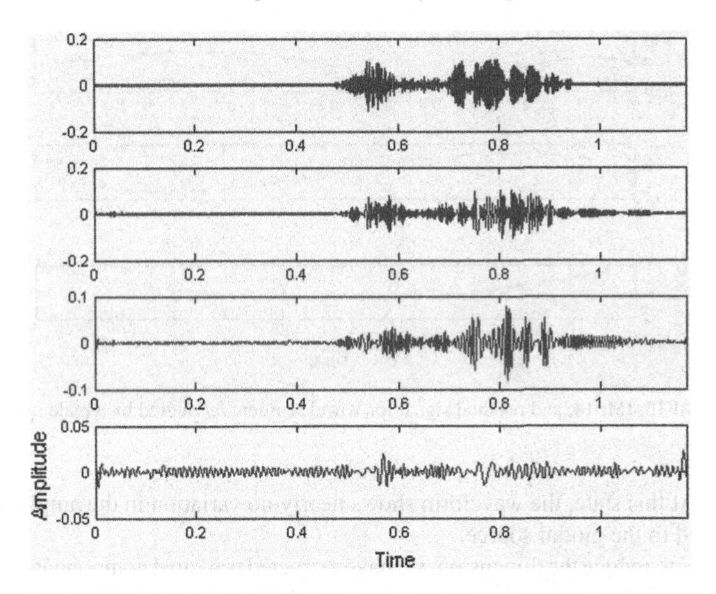

Fig. 9.4 IMF5 to IMF8 for vowel segment /e/ uttered by a male speaker

As mentioned in Sect. 9.5, a total of 125 CVC and VCV tokens are used for LVQ codebook design. Thus, we have obtained 450 vowel segments from which EMD residual is extracted. Figures 9.3, 9.4, 9.5 and 9.6 show the IMFs and residual for the vowel /e/ uttered by a male speaker, where the residual is seen in the last segment of

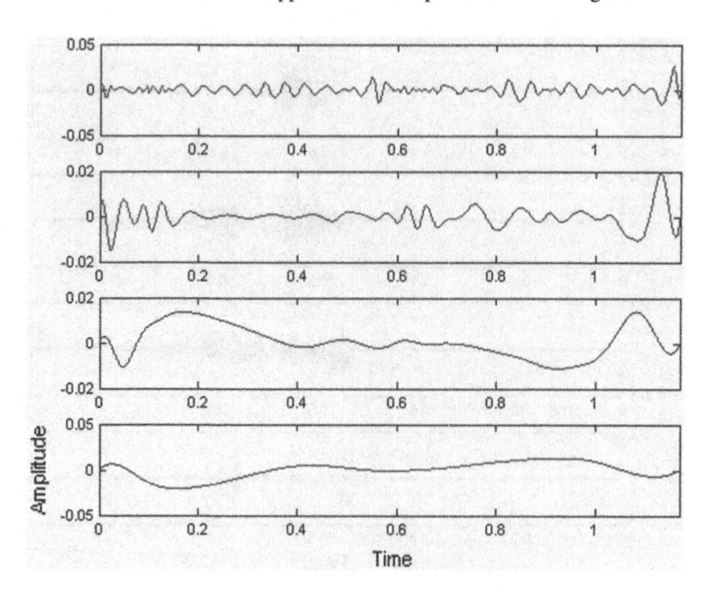

Fig. 9.5 IMF9 to IMF12 for vowel segment /e/ uttered by a male speaker

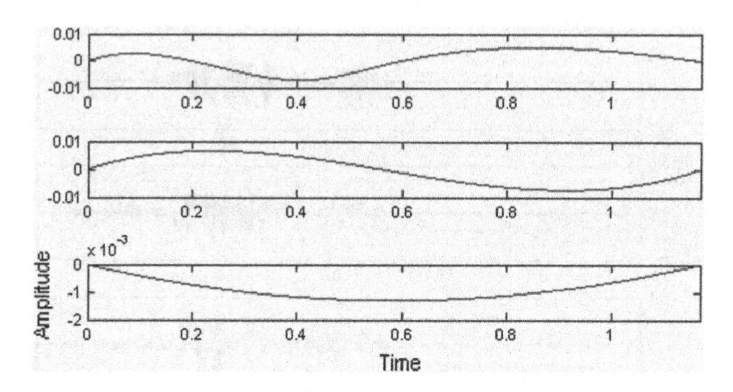

Fig. 9.6 IMF13, IMF14, and residual signal for vowel segment /e/ uttered by a male speaker

Fig. 9.6. At this state, the waveform shows nearly no variation in the amplitude and it is related to the glottal source.

In order to reduce the dimension, we have extracted principal components from the EMD residual signals and these are used to train the LVQ network. The LVQ network thus learns the frequency characteristics of the speaker's glottal source. This trained LVQ network is used for pattern matching with the SOM-segmented vowel, which is obtained from the CVC-type words spoken by the speaker to be identified. The SOM-segmented vowel is passed to the EMD block to get the residual from which principal components are extracted to use in the speaker identification purpose.

Initially, speaker to be identified is asked to utter one CVC-type word containing any of the Assamese vowel and the vowel part is extracted out by the SOM and

Table 9.7 Correct identification of boy speaker-I for EMD residual

S. No	Segmented vowel	Word-I	Word-II	Word-III	Word-IV	Word-V
1	/i/	√	√	√	√	×
2	/e/	×	√	√	√	√
3	/ε/	√	√	√	√	√
4	/a/	√	×	√	√	×
5	/ɑ/	×	√	√	√	√
6	/ɔ/	√	√	√	×	√
7	/o/	√	√	√	×	×
8	/u/	√	√	√	√	√

Table 9.8 Correct identification of girl speaker-I for EMD residual

S. No	Segmented vowel	Word-I	Word-II	Word-III	Word-IV	Word-V
1	/i/	×	√	√	√	√
2	/e/	√	√	√	√	×
3	/ε/	√	√	√	√	√
4	/a/	×	×	√	√	√
5	/ɑ/	×	√	√	√	√
6	/o/	×	√	√	√	×
7	/ɔ/	√	√	√	×	√
8	/u/	√	√	√	√	×

Table 9.9 Speaker identification success rate over all the vowel and word variations using EMD residual

S. No.	Speaker	Success rate (%)
1	Boy speaker 1	94
2	Boy speaker 2	93.2
3	Boy speaker 3	87
4	Boy speaker 4	90
5	Boy speaker 5	80.6
6	Boy speaker 6	90
7	Boy speaker 7	85.2
8	Boy speaker 8	87
9	Boy speaker 9	90
10	Boy speaker 10	80.6
11	Girl speaker 11	95
12	Girl speaker 12	87.8
13	Girl speaker 13	92
14	Girl speaker 14	94
15	Girl speaker 15	85.4
16	Girl speaker 16	86
17	Girl speaker 17	87.8
18	Girl speaker 18	89
19	Girl speaker 19	94
20	Girl speaker 20	89.4
Total	Average	88.9

PNN blocks. Then, it is sent to the to the EMD block to obtain the residual and from the residual signal principal components are extracted which is then presented to the LVQ network to identify the speaker. The correct speaker identification of segmented vowel from five different words containing the vowel in case of boy speaker-I and girl speaker-I is shown in Tables 9.7 and 9.8, respectively. The success rate for all the 20 speakers for different CVC words are summarized in Table 9.9. Overall success rate for the proposed model is found to be 88.9 %. This validates the effectiveness of the proposed approach. The success rate obtained from this hybrid ANN-EMD approach is around 90%, which is somewhat less to contemporary methods which use non-ANN statistical tools. The advantage of this approach is however its ability to learn and use it subsequently, which resembles bioinspired computing.

9.9 Conclusion

An ANN-based prototype model for vowel-based speaker identification is described in this chapter using LP and EMD residual as speaker discriminative feature. The accuracy of the speaker identification directly depends on the vowel segmentation and vowel recognition accuracy of the SOM and PNN, respectively. Here, we have used LP and EMD technique to provide speaker's glottal source information in the form of the residual signal, and it is used to discriminate speaker by the LVQ network. It is observed from the experimental results that the success rate of speaker identification is somewhat less in case of EMD residual as compared to LP residual. But use of EMD in the proposed technique proves the usefulness of this novel technique to obtain speaker discriminative features. A self-sustaining, fully automated mechanism for vowel-based speaker identification model has been successfully integrated in this chapter. Although the speaker identification success rate has not reached 100 % in the experiments carried out throughout this work, there is scope to increase it further using more number of samples to train the LVQ network. Further, hybrid approaches can be attempted by integrating fuzzy systems as part of the framework.

References

1. Miles MJ (1989) Speaker recognition based upon an analysis of vowel sounds and its application to Forensic work, Masters Dissertation, University of Auckland, NewZeland.
2. Kumar R, Ranjan R, Singh SK, Kala R, Shukla A, Tiwari R (2010) Text-dependent multilingual speaker identification for Indian Languages using artificial neural network. In: Proceedings of 3rd international conference on emerging trends in engineering and technology, pp 632–635.
3. Lajish VL, Sunil Kumar RK, Lajish VL, Sunil Kumar RK, Vivek P (2012) Speaker identification using a nonlinear speech model and ANN. Int J Adv Inf Technol 2(5):15–24
4. Qian B, Tang Z, Li Y, Xu L, Zhang Y (2007) Neural network ensemble based on vowel classification for Chinese speaker recognition. In: Proceedings of the 3rd international conference on natural computation, USA, 03.

5. Ranjan R, Singh SK, Shukla A, Tiwari R (2010) Text-dependent multilingual speaker identification for indian languages using artificial neural network. Proceedings of 3rd international conference on emerging trends in engineering and technology. Gwalior, India, pp 632–635
6. Chelali F, Djeradi A, Djeradi R (2011) Speaker identification system based on PLP coefficients and artificial neural network. In: Proceedings of the world congress on engineering, London, p 2.
7. Soria RAB, Cabral EF (1996) Speaker recognition with artificial neural networks and mel-frequency cepstral coefficients correlations. In: Proceedings of European signal processing conference, Italy.
8. Justin J, Vennila I (2011) Performance of speech recognition using artificial neural network and fuzzy logic. Eur J Sci Res 66(1):41–47
9. Yadav R, Mandal D (2011) Optimization of artificial neural network for speaker recognition using particle swarm optimization. Int J Soft Comput Eng 1(3):80–84
10. Hu YH, Hwang JN (2002) Handbook of neural network signal processing., The electrical engineering and applied signal processing seriesCRC Press, USA.
11. Templeton TG, Gullemin BJ (1990) Speaker identification based on vowel sounds using neural networks. In: Proceedings of 3rd international conference on speech science and technology, Australia, pp 280–285.
12. Reynolds DA, Rose RC (1995) Robust text-independent speaker identification using Gaussian mixture speaker models. IEEE Trans Speech Audio Process 3(1):72–83
13. Hasan T, Hansen J H L (2011) Robust speaker recognition in non-stationary room environments based on empirical mode decomposition. In: Proceedings of Interspeech.
14. Hsieh CT, Lai E, Wang YC (2003) Robust speaker identification system based on wavelet transform and Gaussian mixture model. J Inf Sci Eng 19:267–282
15. Ertas F (2001) Feature selection and classification techniques for speaker recognition. J Eng Sci 07(1):47–54
16. Patil V, Joshi S, Rao P (2009) Improving the robustness of phonetic segmentation to accent and style variation with a two-staged approach. Proceedings of Interspeech. Brighton, UK, pp 2543–2546
17. Campbell JP (1997) Speaker recognition: a tutorial. Proc IEEE 85(9):1437–1462
18. Huang NE, Shen Z, Long SR, Wu ML, Shih HH, Zheng Q, Yen NC, Tung CC, Liu HH (1998) The empirical mode decomposition and hilbert spectrum for nonlinear and nonstationary time series analysis. Proc Royal Soc Lond A 454:903–995
19. Fakotakis N, Tsopanoglou A, Kokkinakis G (1991) Text-independent speaker recognition based on vowel spotting. In: Proceedings of 6th international conference on digital processing of signals in communications, Loughborough, pp 272–277.
20. Thcvenaz P, Hiigli H (1995) Usefulness of the LPC-residue in text-independent speaker verification. Speech Commun 17:145–157
21. Radova V, Psutka J (1997) An approach to speaker identification using multiple classifiers. Proceedings of IEEE international conference on acoustics, speech, and signal processing 2:1135–1138
22. Sarma SV, Zue VW (1997) A segment-based speaker verification system using *summit*[1]. In: Proceedings of EUROSPEECH.
23. Mahadeva Prasanna SR, Gupta CS, Yegnanarayana B (2006) Extraction of speaker-specific excitation information from linear prediction residual of speech. Speech Commun 48:1243–1261
24. Espy-Wilson CY, Manocha S, Vishnubhotla S (2006) A new set of features for text-independent speaker identification. In: Proceedings of INTERSPEECH, ISCA.
25. Antal M (2008) Phonetic speaker recognition. In: Proceedings of 7th international conference, COMMUNICATIONS, pp 67–72.
26. Jiahong Y, Mark L (2008) Speaker identification on the SCOTUS corpus. J Acoust Soc Am 123(5):3878
27. Ferras M, Barras C, Gauvain J (2009) Lattice-based MLLR for Speaker Recognition. In: IEEE international conference on acoustics, speech and signal processing, pp 4537–4540.

28. Tzagkarakis C, Mouchtaris A (2010) Robust text-independent speaker identification using short test and training. In: Proceedings of 18th European signal processing conference, Denmark, pp 586–590.
29. Shimada K, Yamamoto K, Nakagawa S (2011) Speaker identification using pseudo pitch synchronized phase information in voiced sound. In: Proceedings of annual summit and conference of Asia pacific signal and information processing association, Xian, China.
30. Pradhan G, Prasanna SRM (2011) Significance of vowel onset point information for speaker verification. Int J Comput CommunTechnol 2(6):60–66
31. Kinnunen T, Kilpelainen T, Franti P (2011) Comparison of clustering algorithms in speaker identification. Available via http://www.cs.joensuu.fi/pages/tkinnu/webpage/pdf/ComparisonClusteringAlgsSpeakerRec.pdf
32. Vuppala AK, Rao KS (2012) Speaker identification under background noise using features extracted from steady vowel regions. Int J Adapt Control Signal Process. doi:10.1002/acs.2357
33. Pati D, Prasanna SRM (2012) Speaker verification using excitation source information. Int J Speech Technol. doi:10.1007/s10772-012-9137-5
34. Rilling G, Flandrin P, Goncalves P (2003) On empirical mode decomposition and its algorithms. In: Proceedings of the 6th IEEE/EURASIP workshop on nonlinear signal and image processing, Italy.
35. Bouzid A, Ellouze N (2007) EMD analysis of speech signal in voiced mode.In: Proceedings of ITRW on non-linear speech processing. France, Paris, pp 112–115.
36. Schlotthauer G, Torres ME, Rufiner HL (2009) Voice fundamental frequency extraction algorithm based on ensemble empirical mode decomposition and entropies. In: Proceedings of the world congress on medical physics and biomedical engineering, Germany.
37. Schlotthauer G, Torres ME, Rufiner HL (2009) A new algorithm for instantaneous F0 speech extraction based on ensemble empirical mode decomposition. In: Proceedings of 17th European signal processing conference.
38. Hasan T, Hasan K (2009) Suppression of residual noise from speech signals using empirical mode decomposition. IEEE Signal Process Lett 16(1):2–5
39. Battista BM, Knapp C, McGee T, Goebel V (2007) Application of the empirical mode decomposition and hilbert-huang transform to seismic reflection data. Geophysics 72:29–37
40. Bullinaria JA (2000) A learning vector quantization algorithm for probabilistic models. Proceedings of EUSIPCO 2:721–724
41. Boersma P, Weenink D Praat: doing phonetics by computer. Available via http://www.fon.hum.uva.nl/praat/
42. Fakotakis N, Tsopanoglou A, Kokkinakis G (1993) A text-independent speaker recognition system based on vowel spotting. Speech Commun 12(1):57–68

Chapter 10
Conclusion

Abstract This chapter summarizes the different contents included in the book. It also provides certain limitations in the practical designs included in the work.

Keywords RNN · SOM · CVC · PNN · LVQ

10.1 Conclusion

The book has been prepared to provide the reader an insight into the development of a soft computing framework for phoneme-based speech recognition. Hence, the treatment stated with a brief introduction about speech and its related mechanism. It also discussed about the available literature. It also discussed about the available literature. Certain multi- and interdisciplinary attributes of speech including the perception model of spoken word recognition is covered in Chap. 2. Details of different neurocomputing tools considered are covered in Chap. 3. It is observed that among the time delay structures, the RNN types receive greater attention. SOM has also been a part of speech processing which combines clustering and classification approaches for the purpose. Since the subject considers Assamese language to be the object of application, its phonemical aspects are highlighted in the Chap. 4. The uniqueness of sounds of Assamese language and their significance in speech processing area have also been briefly covered. The Chap. 5 provides the background literature related to speech recognition which constitutes a critical element of the work. It provides the reader the material necessary to comprehend the subsequent portions of the book. The Chaps. 6–9 are derived from actual experimental work. Initially, a SOM-based technique to segment out the initial phoneme from a CVC-type sample is reported. The work provides experimental results better than DWT-based approaches. Next, a hybrid ANN structure is designed using RNN, PNN, and the proposed SOM-based segmentation method to recognize the constituent phonemes of a CVC word. This work is included in the Chap. 7 which however requires a priori knowledge about the

M. Sarma and K. K. Sarma, *Phoneme-Based Speech Segmentation Using Hybrid Soft Computing Framework*, Studies in Computational Intelligence 550,
DOI: 10.1007/978-81-322-1862-3_10, © Springer India 2014

phoneme count. This limitation is removed by the approach described in Chap. 8. Finally, speaker-specific information extracted using LP residual and EMD residual is used with PNN and LVQ ANN blocks to design a reliable vowel-based speaker identification, where the SOM-based segmentation algorithm is again used to segment out the vowel sound. Thus, two distinct applications of SOM-based segmentation algorithm are described throughout the book.

10.2 Limitation

The work has certain limitations. One of the primarily limitations is that the system needs to be effectively configured to make it suitable for real-time applications. The true perception modeling done by ANN can be an addition which can reduce the training time latency presently observed.

10.3 Future Scope

The future scope of the work can be summarized as below:

- The computational time of the proposed SOM-based segmentation technique has to be reduced by using distributed processing system.
- Improving the success rate of K-means clustering-based number of phoneme determination technique, the spoken word recognition model, has to be extended into multiple phoneme case.
- The spoken word recognition model proposed here is based on Assamese phonemical structure. But most of the Indian languages posses similar type of phonemical structure with some distinct phonemical groups. Therefore, for other Indian languages also, such spoken word recognition model can be developed.
- Including more speakers from various dialects of Assamese language, the vowel segmentation-based speaker identification model has to be improved. A study has to be carried out on basic dialects of Assamese language, so that a proper speaker database can be created which includes all the variations of occurrence of vowel sounds in the language.
- The work can be next extended to include the regional variations of native speakers with continuous inputs so that a complete speaker identification or spoken word recognition system for Assamese language can be designed.

This work is expected to motivate researchers to work in this diversion and contribute significantly.

Index

M. Sarma and K. K. Sarma, *Phoneme-Based Speech Segmentation Using Hybrid Soft Computing Framework*, Studies in Computational Intelligence 550,
DOI: 10.1007/978-81-322-1862-3, © Springer India 2014

Printed in the United States
By Bookmasters